COMPUTER SCIENCE, TECHNOLOGY AND APPLICATIONS

ADVANCED SECURITY ISSUES OF IOT BASED 5G PLUS WIRELESS COMMUNICATION FOR INDUSTRY 4.0

Computer Science, Technology and Applications

Additional books and e-books in this series can be found on Nova's website under the Series tab.

COMPUTER SCIENCE, TECHNOLOGY AND APPLICATIONS

ADVANCED SECURITY ISSUES OF IOT BASED 5G PLUS WIRELESS COMMUNICATION FOR INDUSTRY 4.0

VIJEY THAYANANTHAN
EDITOR

nova
science publishers
New York

NOTICE TO THE READER

The Publisher has taken reasonable care in the preparation of this book, but makes no expressed or implied warranty of any kind and assumes no responsibility for any errors or omissions. No liability is assumed for incidental or consequential damages in connection with or arising out of information contained in this book. The Publisher shall not be liable for any special, consequential, or exemplary damages resulting, in whole or in part, from the readers' use of, or reliance upon, this material. Any parts of this book based on government reports are so indicated and copyright is claimed for those parts to the extent applicable to compilations of such works.

Independent verification should be sought for any data, advice or recommendations contained in this book. In addition, no responsibility is assumed by the Publisher for any injury and/or damage to persons or property arising from any methods, products, instructions, ideas or otherwise contained in this publication.

This publication is designed to provide accurate and authoritative information with regard to the subject matter covered herein. It is sold with the clear understanding that the Publisher is not engaged in rendering legal or any other professional services. If legal or any other expert assistance is required, the services of a competent person should be sought. FROM A DECLARATION OF PARTICIPANTS JOINTLY ADOPTED BY A COMMITTEE OF THE AMERICAN BAR ASSOCIATION AND A COMMITTEE OF PUBLISHERS.

Additional color graphics may be available in the e-book version of this book.

Library of Congress Cataloging-in-Publication Data

ISBN: 978-1-53615-538-9

Published by Nova Science Publishers, Inc. † New York

CONTENTS

PREFACE

Advanced Internet of Things (IoT) based wireless communication has recently received a lot of attention due to a wide range of industry 4.0 applications. This book provides an overview of security issues to secure IoT based 5G and 5G+ applications such as cybersecurity issues of vehicular communication. Security issues of a cyber-physical system (CPS) against the energy efficiency and latency are challenging critical problems in the industry 4.0. Sensors based on IoT can be deployed to minimize these problems in many applications such as driverless vehicles. Despite many advanced security issues, deploying IoT-based 5G/5G+ with dynamic and intelligent security solutions for minimizing energy costs and maximizing security issues of CPS will be one of the aims in this book. Following chapter titles of this book provide quick information of the chapter topics.

The chapter "An Overview of 5G And Beyond" presents emerging 5G technologies and expected requirements for enhancing facilities of wireless communication employed in industry 4.0. Advanced techniques of increasing energy saving and efficiency using small cells and some challenges for managing security issues are also considered. In Chapter 2, the author proposes techniques based on Reed Solomon codes for reliable communication in IoT. Chapter 3 directs researchers and 5G developers to understand the efficient data traffic techniques for IoT-based 5G networks. Chapter 4 "Analysis of Cybersecurity for 5G Based Vehicular

Communication Network" presents the security solutions which enhance the security facilities in the future transport systems. Chapter 5 deals with the cybersecurity issues which protect the industry 4.0 where the CPS automats and enables data exchanges. Chapter 6 "Effective Bandwidth Management For Mitigating Denial of Service (Dos) Attack in SDN" shows the SDN-based proposed model and analysis to improve the security solutions. Despite many E_learning facilities, Chapter 7 allows curriculum developers to establish a secure link for enhancing online and virtual learning facilities within the industry 4.0 where E_learning based on 5G system provides quick access with maximum security. Chapter 8 "Traffic load balancing based on routing algorithms for content-centric networking" shows the traffic management using critical generation rate. Managing traffic load is one of the challenging problems in 5G+ and beyond. In Chapter 9, explanation of security and privacy in location-based services is briefly considered for 5G+ mobile communication. Finally, Chapter 10 "Learning style transformation using intelligent IoT, 5G and beyond" provides the additional features to maintain the lifestyle facilities according to the environmental conditions.

ACRONYMS

Acronym	Meaning
EE	Energy efficiency
EH	Energy harvesting
5G	Fifth generation
IoT	Internet of Things
MIMO	Multi-input and multi-output
LDPC	Low-density parity check
LOS	Line of sight
AWCN	Advanced wireless communication network
IIoT	Industrial IoT
DNA	Deoxyribonucleic acid
MAC	Medium Access Control
KPI	Key performance indicator
SE	Spectral efficiency
BH	Back haul
FH	Front haul
AMMS	Adaptive MIMO mode selection
SDN	Software-defined network
NVF	Network virtual function
FSO	Free space optics
Acronym	**Meaning**

NFP	Network flying platform
CRAN	Cognitive radio access network
AMPS	Advanced Mobile Phone System
TACS	Total Access Communication System
NMT	Nordic Mobile Telephone
CDMA	Code division multiple access
WCDMA	Wide band CDMA
SM	Spatial multiplexing
HSDPA	High-Speed Downlink Packet Access
LTE	Long term evolution
WiMAX	Worldwide Interoperability for Microwave Access
UAB	Unified access and BH
LAA	Licensed assisted access
STC	Space-time coding
MEMS	Micro Electro Mechanical Systems
QSF-IM	Quadrature Space-Frequency Index Modulation
RRM	Radio Resource Management
IT	Information technology
FWI	Fiber wireless integration
DSA	Dynamic Spectrum Access
UWB	Ultra-wideband
SON	Self-Organizing Network
WSN	Wireless sensors network
CDN	Content Distribution Networks
URLLC	Ultra Reliable Low Latency Communication
GCHN	Green cognitive heterogeneous network
NLOS	Non-line-of-sight
RS	Reed-Solomon
DVD	digital video disc
VHDL	Verilog hardware description language
SDD	Soft decision decoding
HDD	Hard decision decoding
Acronym	**Meaning**
DFT	Discrete Fourier Transform

BM	Berlekamp Massey
WB	Welch Berlekamp
GF	Galois Field
SoDeMa	Software-defined multiple access
QoS	Quality of services
AFSA	Artificial Fish Swarm Algorithm
NOMA	Nonorthogonal multiple access
SCMA	Sparse Code Multiple Access
PDMA	Pattern division multiple access
MUMA	Multi-User Multiple Access
SACA	Simulated annealing & clustering algorithm
OPNET	Optimized Network Engineering Tool
DDoS	Distributed denial of services
D2D	Device to device
VCN	Vehicular communication network
CS	Cybersecurity
VCF	Vehicular communication functions
SDMC	Software Defined Mobile Network Control
CA-HCIA	Cost-Aware Hierarchical Cyber Incident Analytics
HTTP	Hyper Text Transfer Protocol
SMTP	Simple Mail Transfer Protocol
DNS	Domain Name Servers
ICMP	Internet Control Message Protocol
IDS	Intrusion detection system
CP-ABE	Ciphertext-policy attribute-based encryption
RSA	Rivest–Shamir–Adleman in cryptography
UCON	Usage control
ICS	Industrial control systems
SD	System Dynamics
RADIUS	Remote Access Dial-In User Service
Li-Fi	Light fidelity
Acronym	**Meaning**
RSU	Road service unit
CPS	Cyber-physical system

CC	Cloud computing
BC	Blockchain
TIPS	Traffic information provision services
APT	Advanced persistent threat
SGA	Stochastic geometry approach
M2M	Machine to machine
ICT	Information communication technology
CPPS	Cyber-Physical production systems
RAMI	Reference architecture model industry 4.0
FoF	Factories of the future
ICPS	Industrial cyber-physical systems
TVSTB	Television set-up-box
RCP	Routing Control Platform
BGP	Border Gateway Protocol
ForCES	Forwarding and Control Element Separation
ASR	Aggregation Services Routers
API	Application programming interface
PH-DM	Port Hopping Based DoS Mitigation
TCAM	Ternary Content Addressable Memory
EMS	E_Learning management system
RFID	Radio frequency identification
CCN	Content-centric networking
SD-STAR	Software Defined Status Aware Routing
SP	Shortest Path
ICN	Information-Centric network
MMRP	Markov Modulated Rate Process
ECCN	Edge-centric computing and CCN
RREQ	Route Request
RREP	Route Response
CCMN	Content-centric mobile networks
Acronym	**Meaning**
ISP	Internet service provider
LBS	Location Based Services
GPS	Global Positioning System

POI	Point-of-interest
BLOT	Bloom Filter Oblivious Transfer
VSN	Vehicular Social Networks
ALBS	Advanced LBS
ISHF	Independent secured hash function
AT	Anonymization techniques
OT	Oblivious Transfer
ONF	Open Networking Foundation
CRCA	Cloaked region construction algorithm
IMSI	International Mobile Subscriber Identity
SDIAN	Software Defined Industrial Automation Network
CPIoT	Cyber-Physical IoT
ICPS	Industrial cyber-physical systems

In: Advanced Security Issues …
Editor: Vijey Thayananthan
ISBN: 978-1-53615-538-9

Chapter 1

AN OVERVIEW OF 5G AND BEYOND

Vijey Thayananthan[1,*] and Surekha Lanka[2]
[1]Department of Computer Science,
King Abdulaziz University, Jeddah, KSA
[2]Department of Information Technology,
Stamford International University, Bangkok, Thailand

ABSTRACT

Allowing 5G requirements in the emerging wireless communication system makes sure that users are enjoying the benefits such as speed, latency, etc. Although emerging 5G technologies provide many facilities such as energy efficiency (EE), requirements of evolving technologies expect more to enhance the features. The requirements of the 5G+ and beyond depend on many factors such as growing security solutions, minimum latency, high EE, etc. In addition, delay and latency created by interactions of a growing number of users with emerging technologies and features such as adding new security issues. The energy consumption and interactions related to security problems will be increasing with many factors; they are new threats, simultaneous services, massive applications, etc. In order to improve these problems, 5G+ will be using novel EE and

* Corresponding Author's E-mail: Thayananthan@live.co.uk.

security models within the flying, floating, fixed and mobile network platforms where air water and land communication may be considered with dynamic spectrum management and femtocells. Finding solutions for increasing EE with femtocells and energy harvesting (EH) and minimizing or removing users' interactions with emerging technologies will be some of the future challenges in the 5G+. This chapter introduces the overview of 5G+ with manageable solutions based on EE and emerging technologies.

Keywords: emerging technologies, 5G, IoT, energy efficiency, 5G+ challenges

INTRODUCTION

Next generation of a 5G system known as 5G+ will involve with efficient channel model incorporated to focus on massive multi-input and multi-output (MIMO) which needs LDPC coding to improve the future 5G environments. Although 5G+ is going to have wired and wireless front haul/back haul infrastructure, the roles of the LDPC will simplify many challenging problems in future 5G/5G+ environments. They are the deployment of small cell, line of sight (LOS) solution, non-LOS point-to-multipoint design, etc. However, future solutions depend on the EE and energy harvesting which will be possible through the efficient design of low-complex LDPC coding. The ambition of service provides that future communication systems should be available and reliable for accessing all services whenever users need. The 5G is officially available from 2020 to provide maximum benefits, but still, some public services are expecting a lot of technological challenges against cost which relies on the energy consumptions. Although energy experts found many solutions to minimize the energy cost, implementation of real applications in public services increases the cost. Basic public services such as medical, academic, etc. are very expensive all over the world.

A lifetime of the advanced technologies and storage devices is being extended but the competition of the lifetime depended on the EE is always challengeable with evolving technology and requirements of future systems and users. This chapter investigates the benefits of emerging 5G technologies and advanced 5G wireless communication network (AWCN) architecture such 5G+ from the perspectives of some promising EE techniques. Although AWCN merges with IoT and industrial IoT (IIoT), challenges of EE and techniques of increasing EE are going to be the continuous future problem. To improve the EE, Energy harvesting technologies are being investigated using Micro Electro Mechanical Systems, advanced materials, low complex algorithms based on manifold techniques, etc. Also, the natural energy sources such as solar energy are converted into electric power not only support to reduce the energy cost but also improve the EE.

Security issues with the scent authentication may be introduced and developed in the 5G+ schemes because a growing number of interactions needs to be secured when many users are simultaneously accessing the service. Although fingerprint and image are the basic authentication approaches, the scent of the DNA could be recorded in audio formate and used during the investigation as a 3rd authentication processing.

EMERGING 5G TECHNOLOGIES

Integrating 5G with emerging technologies for spectrum management is one of the targets in smart solutions. Emerging 5G communication technologies provide better 5G network facilities such as capacities improvement against the growing number of users and network traffic. Further network capacities depend on not only the future 5G architectures but also efficient ways of handling network. Again priorities and dynamic, intelligent management take place to improve network capacities when we need. Emerging 5G communication technologies allow the researcher to improve the ambitious key performance indicators (KPIs). In the future emerging 5G, KPIs depend on spectrum sharing, multiple accessing, multiplexing, small cells, etc. Spectrum sharing is one of the techniques for improving spectrum efficiency (SE), and it deals with the allocation of an unprecedented amount of spectrum that could be used for unlicensed or shared services. Although many possibilities which allow researchers to improve the SE, ways of increasing SE depends on the efficient technologies and the design of the system which influences many complex infrastructures. Following ways may be possible to increase the SE, they are intelligent approaches of spectrum handling, systems management with priorities, etc. Although emerging technologies are appropriate, developing solutions are the most important with low cost and complexities despite many theories which provide both maximum security and energy efficiency, emerging 5G communication and its technology motivate us to improve both in future applications.

As shown in Figure 1, tactile Internet depends on the emerging 5G communication technologies and BH/FH solutions. Although IoE and IoT are the main pillars of the tactile Internet, smart BH and FH solutions enhance the technical requirements [12].

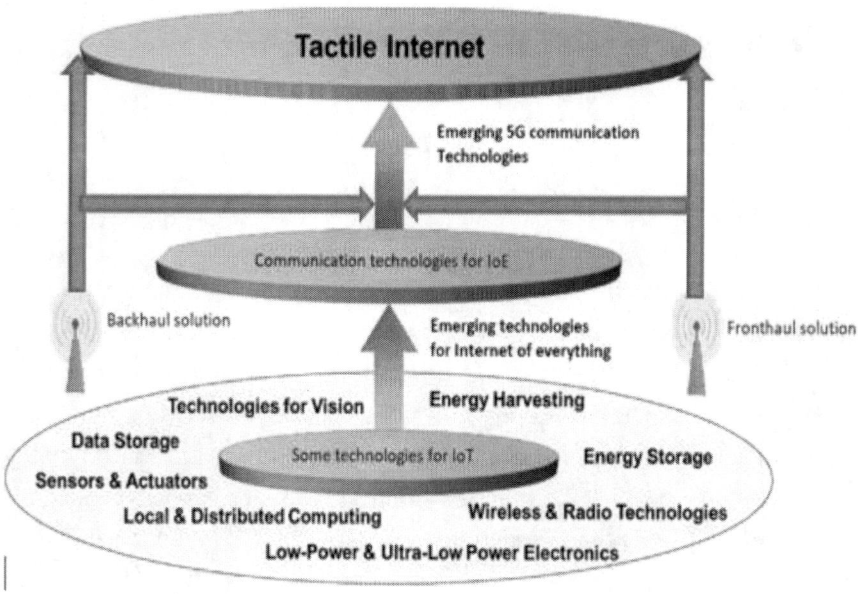

Figure 1. Emerging technologies and BH/FH solutions for Tactile Internet.

IoT Based 5G and 5G Plus

The IoT has extended the scope of services in many public and private sectors within the industry 4.0 where 5G and 5G+ will play their roles when new 5G infrastructure employs the IoT. The IoT will soon pervade all applications involved in our life, technologies merging with services, etc. Massive IoT communications are one of the challenges in 5G+ communication. Evolution of IoT depends on the 4 layers architectures which are services, data, network, and sensors. Although standardization of IoT is not defined, 5G+ will enhance the future Internet facilities such as security solutions of the architecture stronger. As shown in Figure 2, mission-critical communication based on 5G+ will enhance the capacity and limitation of the IoT and Tactile Internet services and facilities.

The future Internet influenced to mission-critical communication will be a promising paradigm to handle massive IoT devices and their connections in 5G, 5G+ and beyond. Spectrum sharing solutions based on IoT provide 2 approaches they are IoT technologies deployed within licensed and unlicensed cellular spectrum. These days, end devices in IoT spectrum sharing are very expensive and consuming more energy. Therefore, the design of IoT spectrum sharing should be efficient to improve the benefits which fit to deploy 5G and beyond. Further, 5G+ will be expecting efficient design such as EE, limited signal processing, etc.

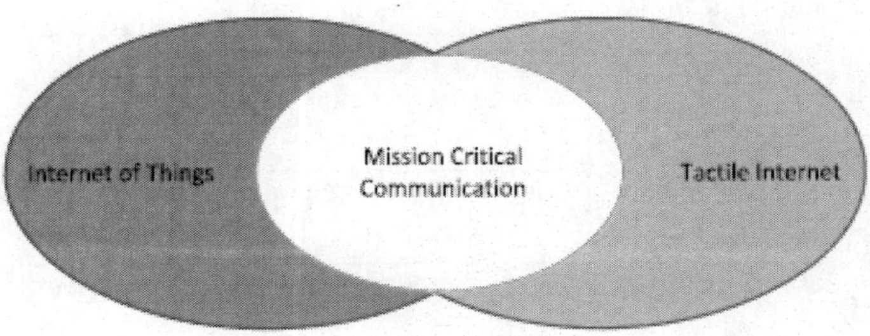

Figure 2. Mission-critical communication for IoT and Tactile Internet [10].

Emerging 5G Communication Systems

Although we handle many emerging 5G communication technologies, we define that technology allows us to enhance the characteristics and performance of future communication during the integrations two or more schemes. Spectrum efficiency is one of the requirements to improve emerging 5G technologies. As shown in Table 1, we can provide future smart BH and FH solutions which provide not only the improvement details of spectrum management but also users' flexibilities when multiple devices deal with simultaneous communication links and connections. In the emerging 5G communication strategies, overall network SE influences with femto cellular approach and resources. Here, densification is one of the approaches to improve spectrum management.

Also, Table 1. shows the BH data streams which incorporate with adaptive MIMO mode selection (AMMS) to improve 5G transmission processing.

Table 1. Emerging 5G technologies for smart solutions

Emerging 5G technologies	Smart BH and FH solutions
Massive MIMO and AMMS	Lowering access delay and avoiding traffic congestions intelligently
Hybrid existing wireless - mmWave communication	Intelligent dynamic spectrum sharing & Interference mitigation
Integrated mmWave-microWave communication	Enhances the basic requirements and provides high-speed communication
Integration of NVF and SDN	Improve the throughput and enable the compatibility through the SDN
Software-defined multiple access	Improve the spectrum sensing and dynamic spectrum management for avoiding traffic congestion

REQUIREMENTS FOR 5G+ ARCHITECTURE

Advances in free space optics (FSO) mentioned in [1] motivate 5G+ researchers to consider new features and requirements for 5G+ architecture. Networked flying platforms (NFPs) such as drones will enhance the communication services by 2030 when 5G+ becomes matured and merged technology with all real communication systems. FSO-based fronthaul/back haul framework is introduced as an application of NFP to improve network throughput [2]. Expected 5G+ requirements will be the users' satisfaction which includes the quick availability and easy deployability offered by FSO [4].

When we deploy high BH and FH capacity as a smart solution, centralized architecture in the emerging 5G technologies allows us to have better joint processing and higher SE. When we use distributed architecture, we can deploy a flexible network with lower SE. In this architecture, managing interference will be difficult. Although the flexible network is better for connecting all application adaptively, different BH/FH capability will require spectrum sharing to achieve sufficient spectrum availability in some scenarios. Figure 3 shows the example of a 5G+ design using recent schemes such as small cells.

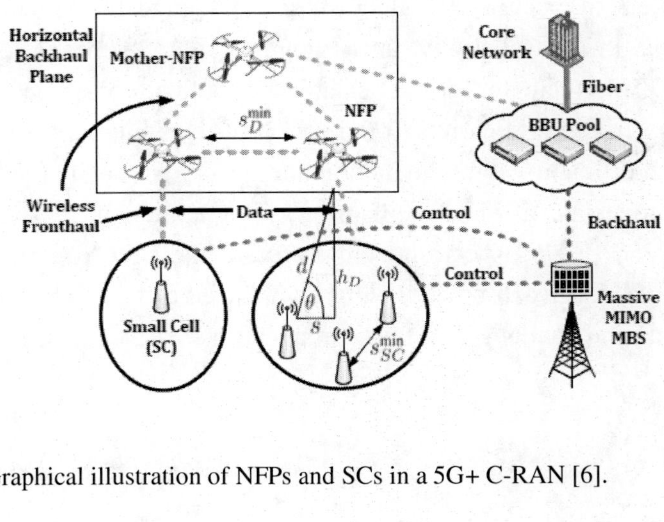

Figure 3. Graphical illustration of NFPs and SCs in a 5G+ C-RAN [6].

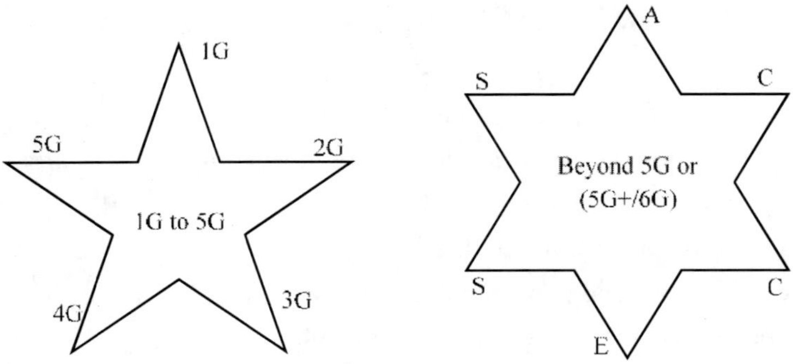

Figure 4. Evolution of technologies (a) 1G to 5G and (b) Beyond 5G.

In all last 5 generations depicted in Figure 4a, continuous improvement of each generation (G) is briefly introduced in Table 2. Each generation focused on the rate being increased with evolving technologies. Generations 5G+ and 6G are also going to focus on increasing the rate as well EE which is one of the most important requirements in the future. Although it helps us to reduce the cost of the massive services, specifically EE safe the lifetime of the devices.

In order to organize the 6G, Figure 4b introduces the expected requirements with the acronym "ACCESS." Despite many expectations, ACCESS encourages researchers, service providers and users to debate and discuss the future requirements of the 5G+ or beyond.

In Table 3, the accuracy of latency is still challengeable with the specific applications. In controllability, an acronym of RAMS means Reliability, Availability, Maintainability, and Scalability should be considered to control the 5G+/6G from the security. Computation of dynamic bandwidth should be managed efficiently with the natural energy sources such as solar power which improves the EE. Despite many ways of achieving SE, dynamically managing and sharing spectrum will increases the SE. As far as SDN is concerned, 5G+/6G will be deploying SDN based technologies to improve the demands and challenges of future applications.

Table 2. Evolution of mobile technologies

G	Evolution of technologies	Year	Rate
1G	AMPS, TACS, NMT 2.4kbps Analog telephones	1980-1990	100bps to 10kbps
2G	64kbps digital, 2G voice, SMS, kb Internet	1990-1995	10kbps to 100kbps
3G	CDMA2000, WCDMA, HSDPA, LTE, etc. 3G voice, SMS, Mb Internet	1995-2010	100kbps to 100Mbps
4G	LTE advanced, WiMAX, etc. 4G voice, Mb Internet	2010-2020	100Mbps to 10Gbps
5G	Gb Internet, Mobile broadband, Massive connectivities, low latency, Ultra reliable communication, etc.	2020-2030	10Gbps to 100Gbps
5G+/6G	FSO, NFP	2030-2050	>100Gbps

Table 3. Expected requirements of 5G+ and 6G

Short form of 5G+/6G features	Features of 5G+/6G
A	The accuracy of latency, capacity, etc.
C	Controllability of RAMS
C	Computation of dynamic bandwidth efficiency
E	EE based on ES and EH created from natural resources (solar)
S	SE with dynamic spectrum sharing
S	SDN based technologies (SoDeMa)

SMALL CELLS AND TECHNIQUES OF INCREASING EE

The upcoming formations pertaining to the 5G domain can anticipate the innovation and extensive integration of the small cell. These cells are mobile telephone nodes that require minimal power consumption ensuring optimum utilization of the frequency inside the finite region while enhancing the bandwidth efficiency. The reinforcement of bandwidth efficiency is barring any further increment of the range results in the rapid expansion of condensed networks amongst numerous interconnected IoT devices [21, 22].

As shown in Figure 5, the small cell will strengthen the remote system and increment data rates to clients. Here, an access point (AP), long term evolution (LTE), television set-up-box (TVSTB) are considered.

The concept of Femtocell is rejuvenating by telephone operators who have represented until now the continuation of more primitive technologies like 3G and 4G. With the assimilation of Technology 5G, Engineers are acknowledging the deploying of the femtocell indoors. The utilization of unlicensed spectrum for LTE (LTE-U) and licensed assisted access (LAA) using band proposed for Wi-Fi systems and indoor and out-of-doors. Until today, these sort of deployments were restricted to places that considerable numbers of endorsers like airplane terminals, malls, brandishing occasions. The high data utilization request in these spots requires an explicit sort of infrastructure to enhance the efficiency of the network system, strengthening it using small cells.

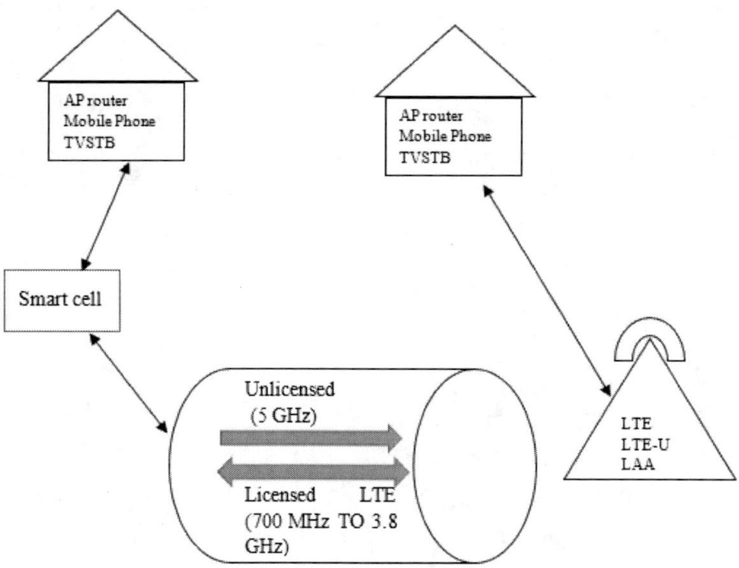

Figure 5. A small cell in some selected clients.

Survey Results on Small Cell

The small cells have a greater capacity of the bandwidth of approximately 6GHZ, which would result in the frequency to become more vulnerable to barriers while propagating. In order to tackle the issue, a higher number of cells must be deployed to ensure optimal coverage. According to the case study conducted by IHS Markit Company among service providers plan boots which were deployed both interior and exterior area skirts. The list of results according to the survey is an in-building limited elaboration; large-scale networks organize resource offload, out-building limited elaboration, non-expandability of macrocell setup, streamlining of high data utilization zones, open building coverage elaboration and-building inclusion streamlining [24].

Small Cell Advantages

Small cells are a pivotal aspect of 5G since it has the ability to enhance the volume, frequency and connectivity range of the network. Furthermore, these cells are crucial to establishing 5G as a functional, rational and cost-effective replacement to 4G. Despite being confined to a certain area, femtocells can be arranged and assembled to provide absolute cellular connectivity throughout the geographical area. These results in the clients to be connected uninterruptedly from one receiver to the other without losing any network packet. The salient features of small cells include the following: greater capacity, wider area coverage, and enhanced bandwidth. Higher capacity would result in more compact cells' connection amongst the widely distributed area [23].

Challenges of EE in 5G networks can be analyzed using the following modifications.

- Sampling techniques
- Intelligent routing protocols
- Mobility challenges in different environmental conditions

- Cognitive radio technology influenced by next-generation networks
- Service differentiation and Gigabit wireless connectivity with EE
- Analysis of EE with Fragmented Spectrum and Carrier aggregation

In a 5G network, femtocells (small cells) can be considered to increase the EE than the large cells because its coverage is very high. Following concepts are considered when femtocells are employed in the 5G network [11].

- *Femtocell design challenges for EE:* It can be used either fixed or mobile places, but design challenge is maximum coverage within a few square meters
- *Efficient and less complex components:* Network components which connect with femtocells and Routing protocols must be considered
- *Energy efficient management:* Minimize interference, delays, and congestions between the cells
- *Maximizing capacities:* Maximum number of users within a few square meters is considered with external environmental conditions
- *Energy efficient architecture:* Design of energy efficient routing protocols and relays in the 5G network system should be considered with efficient technology which includes hardware and software.

Following three key areas can be considered to achieve the energy saving

1. More spectrums must be pressed into service: New spectrum will be expected with the increasing demand for using large capacities and coverage. Using more new spectrum, energy consumption will be increased in 5G development. So, the amount of spectrum should be designed efficiently with energy efficient components.
2. Networks will become much denser with many more cells: Many more BS is another important development in the future 5G, which

is flexible to merge with any networks. But, energy consumption will be very high when many network components are active in each cell. Network components connected with the physical layer should be considered with EE design.

3. Raising the overall performance of networks: Radio and 5G access technologies should be designed with energy efficient physical layer which provides services to upper layers including the network layer.

Emerging 5G technologies need maximum EE studied in this section and harvesting or saving energy is also important to increase EE.

EMERGING 5G CHALLENGES AND OPPORTUNITIES

Table 4 shows the emerging 5G technologies which provide some smart approaches and challenges.

Although 5G+ expects to solve many challenges, emerging 5G technologies allow us to note quick opportunities, challenges, and solutions for illustrating some applications.

Opportunities

Although many hottest trends in communication technology which uses opportunities to transform the future using dynamic solutions, few intelligent approaches increase the opportunities for improving the applications of the communication. Ultra-compact devices emerged with 5G communication, and network technologies are able to handle in the multi-polarization environment and multiband. Here, many opportunities can be raised to improve the research of dynamic spectrum; they are such as hybrid (solar and wireless power) and EH-MEMS. Also, these devices are able to function in a wide dynamic range with high-frequency operations.

Challenges

According to future applications such as driverless vehicles depend on many challenges such as intelligent based priority and dynamic solutions. These challenges have to be considered with emerging 5G communication technologies. Although this article introduces some challenges, we will be facing many evolving problems such as security and energy efficiencies with a growing number of users and all types of communication devices.

Table 4. Emerging 5G technologies and challenges

Emerging 5G technologies	Smart approaches with BH and/or FH	Challenges
Massive MIMO and AMMS	SM, STC, etc. with the intelligent	low access delay and avoiding traffic congestions
Intelligent dynamic spectrum with mmWave communication	UAB and multi-hop BH through the better design	Dynamic controlling, accessing intelligently & interference mitigation
Integrated mmWave-microWave communication	EH-MEMS, RF-MEMS & URLLC-aware frame structure	Energy improvement, low-cost, high-speed communication with the lowest latency
Integration of NVF and SDN	Software-defined H-CRAN system	Flexible and capable core networks with dynamic network conditions
Software-defined multiple access	Inband, outbound and All (5G) cellular communication devices	Tactile Internet with flexibility, scalability, efficiency, etc.

Solutions

Tactile Internet requires some improvements in the design and solutions which depend on the significant reduction in latency. Although the concepts of the Tactile Internet are defined in [12], future challenges of design and solutions depend on the emerging technologies of 5G. It allows remote users to improve real-time control through the ultra-reliable and ultra-responsive network connectivity. Despite many solutions from the upper layers, some hardware solutions depend on MEMS (Micro Electro Mechanical Systems). In the future Internet, a smart solution of minimizing energy cost depends on energy harvesting (EH-MEMS), and radio frequencies (RF-MEMS) intend to use in the 5G applications and beyond.

LITERATURE REVIEW

According to [3], applying 5G technology with back haul at the Radio Resource Management (RRM) and Medium Access Control (MAC) layers enhances the KPI of the current 5G requirements. This enhancement leads future researchers to design the 5+ system and architecture. Building extended Dynamic Spectrum Access (eDSA) with back haul evolution will enhance the 5G+ KPIs.

Future 5G+ networks expect to have a simple and manageable traffic model which should be able to measure the various properties of real-time traffic with maximum accuracy and minimum initial conditions [7].

Fiber wireless integration (FWI) will be one of the technological options in 5G+ cellular communication networks [8]. Further, FWI supports to develop future broadband and ultra-wideband (UWB) wireless system with minimum interference.

Self-Organizing Network (SON) may be used as one of the challenges to improving information security when NFP provides the transmission over the wireless network. Here, the SON's functionalities allow securing coverage expansion or capacity enhancement [5].

According to [14], the use of sensor and application of IoT are growing with the evolving technology such as 5G, 5G+, etc. Thus, establishing Wireless Sensor Network (WSN) with current and future wireless communication enhances the facilities with low-cost. Wherever we deploy the IoT, WSN and its application will provide the necessary solutions such as energy efficiency which increases the lifetime of the IoT devices.

Integration of both energy harvesting and serving (lifetime ES capacity of storage devices) will allow us to develop the energy efficient system for minimizing the energy cost of the future 5G/5G+ wireless applications such as e_Health applications over a wireless network in smart cities [15].

Flexible CDN (fCDN) architecture is one of the technological concepts expected to design in the 5G+ networks [16]. Thus, the authors proposed fCDN for improving 2 problems suffered in the existing CDN. They are DNS-based mapping and content reflection eliminated through the proposed algorithms such as Greedy and Swing [16].

According to [13], the thermoelectric conversion for harvesting energy requires materials with high electrical conductivity. Despite many natural energy sources such as heat, light, etc., novel energy harvesting materials and nanostructures can be considered to employ in future 5G/5G+ wireless applications.

As far as a smart solution is concerned, hybrid wired-wireless BH solutions such as connectivity and capacity provide some benefits and improvements to heterogeneous ultra-dense network [20].

Article [17] explains the integrated mmWave-microWave communications and introduces the Ultra Reliable Low Latency Communication (URLLC)-aware frame structure. Although mmWave communication alone may not provide enough support, integrated version improves the wireless communication future challenges. Due to the spectrum scarcity, existing wireless networks could not allow researchers to enhance the requirements of future services. But this integrated version not only enhance the basic requirements but also it provides high-speed communication.

A green cognitive heterogeneous network (GCHN) is one of the future challenges that gets support from the decentralized settings with the uplink spectrum and BH energy allocation. In this research, wireless traffic is forwarded to the gateway via BH link with two small base stations. Using a renewable power supply commonality in the proposed model authors achieved better GCHN [18].

According to [19], smart BH solution depends on the millimeter wave (mmW) technologies because this emerging approach allows researchers to enhance the communication requirements of the outdoor urban small cells and 5G cellular spectrum. In this solution, the non-line-of-sight (NLOS) operation might be the affected the overall performance. Despite the NLOS, the reinforcement learning algorithm can be considered with BH. This algorithm not only improves the reliability of a mmW NLOS small cell BH systems but also smart solutions of a 5G cellular system based on propagation by diffraction.

USE CASES OF ADVANCED SECURITY ISSUES

Although use cases of mobile Internet are going to enhance the future 5G systems used in industry 4.0, security issues are facing many challenges because potential risks and threats are growing with evolving technology. When potential systems merge with the 5G+ technology, users expect to have flexible systems without any threats (0 threats) or attacks. Energy saving is another key challenge which needs to be harvested from the potential resources rather than traditional or conventional sources. Most of the potential gadgets and wearable devices depend on the low power which is possible through the vibrations of the human body. Harvesting energy and providing security solutions with autonomous processing, intelligent approach and decision making based on priorities are some of the future requirements in the 5G+. Despite many use cases, handling Internets with all environmental conditions will be the major challenge in the future 5G+based use cases.

All sectors somehow involved with 5G systems directly or indirectly for managing their daily activities. Use cases are still growing with many potential applications dominated by the Internets which are available for fixed and mobile users. Future service providers of mobile communication expect to have a number of use cases with more features which are flexible to use the Internet anywhere we want. It means that the Internet should be able to connect and accessible in all environmental conditions. This is the place; we have to bring new challenges which will certainly go beyond the 5G features and requirements. In this chapter, we have introduced the new feature of 5G+ which will fulfill the future use cases of the 5G and its new applications. Despite many new applications, we have to focus on developing integrated and flexible to support the demands of next-generation applications. For instance, driverless vehicles expect maximum security with low energy consumptions through the future Tactile Internet applications.

Use cases of advanced authentication depend on many factors such as gait based biometric authentication. Thus, future requirements of IoT based 5G+/6G considered as "ACCESS" will allow the security developers to improve the future security solutions which not only support the gait based biometric authentication but also cybersecurity issues.

CONCLUSION AND FUTURE DIRECTIONS

In the following 2 subsections, we provide a summary of the conclusion studied in this chapter and future directions to open new research challenges.

Conclusion

We have considered the emerging 5G technologies for improving features of evolving technologies such as IoT based 5G and 5G+. This chapter provides some expected requirements of 5G+ architecture which helps us to enhance future communication services. Emerging 5G technologies will be leading to finalize the 5G+ and requirements of 6G which is expecting some predictions such as intelligent priority services according to the situation. We considered some opportunities, challenges, and solutions which influence smart approaches with back haul and/or front haul technologies. They are Ultra-dense of small cell networks, an optical wireless interface for configuring hardware, unconditional connectivity and constant speed of continuous accessing facilities. Although some use cases are noted, we have considered a few potential applications and future challenges.

Future Directions

The interactions of 5G with 50 billion devices by 2020 will be increasing when users are browsing the Internet continuously. Some devices are automated for 24 hours and 7 days to monitor the continuous progress of services which depend on the 5G or 5G+ requirements. Challenge is managing future requirements for improving continuous services and reducing interactions with the available spectrum sharing facilities. Although some interactions are important to enable the system according to the schedule and applications, sometimes they interrupt the whole systems.

Many challenges are available, but few will be feasible in real applications because some of them are practically difficult at this stage they are such as molecular communication, DNA based security, gait based biometric security issues, etc. With the 5G+ features, E_peace and E_government, etc. are also possible in the future. Further, network-based on molecular communication will help us to predict the medical symptoms transferred from the body to 5G networks. Strong security is available, but still, illegal activities related to cyber attacks are growing in modern countries where DNA solution and 5G+ secure network system may be considered.

REFERENCES

[1] Alzenad, Mohamed, Muhammad Z. Shakir, Halim Yanikomeroglu, and Mohamed-Slim Alouini. "FSO-based vertical back haul/fronthaul framework for 5G+ wireless networks." *IEEE Communications Magazine* 56, no. 1 (2018): 218-224.

[2] Gu, Zhiqun, Jiawei Zhang, and Yuefeng Ji. "Topology Optimization for FSO-Based Fronthaul/Backhaul in 5G+ Wireless Networks." In *2018 IEEE International Conference on Communications Workshops (ICC Workshops)*, pp. 1-6. IEEE, 2018.

[3] Frascolla, Valerio, Bismark Okyere, Shahid Mumtaz, Jonathan Rodriguez, Kazi Saidul Huq, Andreas Georgakopoulos, Evangelos

Kosmatos et al. "Breaking the Access Technologies Silos by Enhancing MAC and RRM in 5G+ Networks." In *2018 European Conference on Networks and Communications (EuCNC)*, pp. 1-9. IEEE, 2018.

[4] Gunathilake, Nilupulee Anuradha, and Muhammd Zeeshan Shakir. "Empirical performance evaluation of FSO availability under different weather conditions." In *8th International Conference on the Network of the Future (NOF)*, vol. 978, no. 1, pp. 156-8. 2017.

[5] Ahmadi, Hamed, Konstantinos Katzis, and Muhammad Zeeshan Shakir. "A novel airborne self-organizing architecture for 5G+ networks." In *Vehicular Technology Conference (VTC-Fall), 2017 IEEE 86th*, pp. 1-5. IEEE, 2017.

[6] Shah, Syed AW, Tamer Khattab, Muhammad Zeeshan Shakir, and Mazen O. Hasna. "Association of networked flying platforms with small cells for network-centric 5G+ C-RAN." In *Personal, Indoor, and Mobile Radio Communications (PIMRC), 2017 IEEE 28th Annual International Symposium on*, pp. 1-7. IEEE, 2017.

[7] Mirahsan, Meisam, Ziyang Wang, Rainer Schoenen, Halim Yanikomeroglu, and Marc St-Hilaire. "Unified and non-parameterized statistical modeling of temporal and spatial traffic heterogeneity in wireless cellular networks." In *Communications Workshops (ICC), 2014 IEEE International Conference on*, pp. 55-60. IEEE, 2014.

[8] Yu, Jianjun, Xinying Li, and Wen Zhou. "Tutorial: Broadband fiber-wireless integration for 5G+ communication." *APL Photonics* 3, no. 11 (2018): 111101.

[9] Hussain, Faisal, Md Yeakub Hassan, Md Sakhawat Hossen, and Salimur Choudhury. "System Capacity Maximization with Efficient Resource Allocation Algorithms in D2D Communication." *IEEE Access* (2018).

[10] Zhang, Qi, and Frank HP Fitzek. "Mission critical iot communication in 5g." In *Future Access Enablers of Ubiquitous and Intelligent Infrastructures*, pp. 35-41. Springer, Cham, 2015.

[11] Alzahrani, Ahmed, Vijey Thayananthan, and Muhammad Shuaib Qureshi. "Relevance of Energy Efficiency Gain in Massive MIMO

Wireless Network." *International Journal of Advanced Computer Science and Applications* 8, no. 12 (2017): 351-356.

[12] Iannacci, Jacopo. "Internet of things (IoT); internet of everything (IoE); tactile internet; 5G–A (not so evanescent) unifying vision empowered by EH-MEMS (energy harvesting MEMS) and RF-MEMS (radio frequency MEMS)." *Sensors and Actuators A: Physical* (2018).

[13] Akinaga, Hiroyuki, Hiroyuki Fujita, Masaki Mizuguchi, and Takao Mori. "Focus on advanced materials for energy harvesting: prospects and approaches of energy harvesting technologies." *Science and Technology of Advanced Materials* 19, no. 1 (2018): 543.

[14] Sharma, Pradip Kumar, Young-Sik Jeong, and Jong Hyuk Park. "EH-HL: Effective Communication Model by Integrated EH-WSN and Hybrid LiFi/WiFi for IoT." *IEEE Internet of Things Journal* (2018).

[15] Weddell, Alex S., and Michele Magno. "Energy Harvesting for Smart City Applications." In *2018 International Symposium on Power Electronics, Electrical Drives, Automation and Motion (SPEEDAM)*, pp. 111-117. IEEE, 2018.

[16] Al-Naday, Mays, Martin J. Reed, Janne Riihijärvi, Dirk Trossen, Nikolaos Thomos, and Mohammed Al-Khalidi. *"fCDN: A Flexible and Efficient CDN Infrastructure without DNS Redirection or Content Reflection."* arXiv preprint arXiv:1803.00876 (2018).

[17] Semiari, O., Saad, W., Bennis, M. and Debbah, M., 2018. *Integrated Millimeter Wave and Sub-6 GHz Wireless Networks: A Roadmap for Ultra-Reliable Low-Latency Communications.* arXiv preprint arXiv:1802.03837.

[18] Li, D., Deng, P., Xu, Y., Gao, L. and Zhang, G., 2018. *Joint Access Spectrum and Backhaul Energy Allocation for Green Cognitive Heterogeneous Networks.* IEEE Access.

[19] Malila, B., Falowo, O. and Ventura, N., 2018. Intelligent NLOS Backhaul for 5G Small Cells. *IEEE Communications Letters*, 22(1), pp. 189-192.

[20] López, Onel LA, Hirley Alves, Richard Demo Souza, and Matti Latva-aho. *Hybrid Wired-Wireless Backhaul Solutions for Heterogeneous Ultra-Dense Networks.* arXiv preprint arXiv: 1803.02236 (2018).

[21] Teldat Group, 2018. *"5G: the expansion of small cells"* Teldat R&D Department, https://www.teldat.com/blog/en/5g-network-small-cell-roll-out-femtocell-iot-28ghz-38ghz-frequency/.

[22] Jundhare, Mayuri D., and A. V. Kulkarni. "An overview and current development of femtocells in 5G technology." In *Advances in Electronics, Communication, and Computer Technology (ICAECCT), 2016 IEEE International Conference on*, pp. 204-209. IEEE, 2016.

[23] Small Cell Forum Ltd 2007-2018. "Why small cells are important for 5G," *Small cell definition*, https://www.smallcellforum.org/what-is-a-small-cell/.

[24] Campbell, Karen, Jim Diffley, Bob Flanagan, Bill Morelli, Brendan O'Neil, and Francis Sideco. The 5G economy: How 5G technology will contribute to the global economy. *IHS Economics and IHS Technology* (2017).

In: Advanced Security Issues …
Editor: Vijey Thayananthan
ISBN: 978-1-53615-538-9

Chapter 2

REED SOLOMON CODES FOR RELIABLE COMMUNICATION IN INTERNET OF THINGS

Ali Mahmudi
Software Engineering Department,
National Institute of Technology (ITN), Malang, Indonesia

ABSTRACT

In the networking field, Internet of Things (IoT) is the current state of the art in the nowadays Information of Technology era. The networking may be defined as an external or internal network. The backbone of the IoT is the internet connections. The IoT connects various objects to the Internet so that they can communicate and exchange billions of data and information among various devices and services. They may be remotely controlled from a distant area. As IoT systems will be open and available everywhere, a number of security issue may arise. One issue that remains open in the IoT technology is security and privacy issues. Because of this security issue, the communications among many different devices powered by IoT could not be said as a reliable technology.

Because of this, the security of the IoT systems can be enhanced by adding an error correction scheme both in the communication channel as well as the data store. By introducing the error correction scheme, the risks may be reduced to an acceptable level, and the security could be enhanced.

A Reed-Solomon (RS) code is one of many error control coding schemes that was firstly introduced by Reed and Solomon in 1960. This code has been used in various applications, such as CD-ROMs, space communications, DVD technology, digital TV and much more.

Here, the RS code is discussed in detail. It raised the issue of RS decoding scheme using the Welch Berlekamp algorithm. It presents the implementation of the Welch Berlekamp algorithm for the RS decoder in detail. The VHDL implementation VHDL for Hard Decision Decoding using the Welch Berlekamp algorithm is also presented. Without loss of generality, the Hard Decision Decoding for RS(255, 239) over GF(2^8) is developed using VHDL.

Keywords: Internet of Things, error control codes, Reed Solomon codes, VHDL

1. Introduction

Internet of Things was firstly introduced by Kevin Ashton in 1999. Fifteen years ago, there is not yet a global consensus on the definition of IoT. In general, the IoT is defined as the ability to connect various objects using internet connection as the backbone and allows them to interact with other objects, or with other intelligent computing equipment via the internet network. IoT has been used in various applications in many aspects of human life [1].

Widespread use of IoT technology in modern life turns making human life to be much more comfortable. IoT has affected human life nowadays, from the domestic home appliances up to smart cars. In business point of view, IoT has been used in decision making to increase production as well as the quality of production, to oversee the distribution of goods, to prevent counterfeiting, to shorten the waiting time of unavailability goods in retail markets, in supply chain management, and so on.

IoT has also been used in the public health sector. IoT has been used in medical applications, for example, for monitoring glucose in a diabetic patient. A doctor can receive data in real time to monitor the patient's condition and to manage the medicine of a patient from home by using IoT application as a backbone. Thus, looking after and monitoring a patient becomes much easier to be done. IoT has also been adopted in smart home applications which makes it possible to arrange and control some home appliances remotely from a distant area [1].

IoT consists of different devices. Some technologies involved is RFID as an object and location identifier, WSN (Wireless Sensor Network), cloud computing and so on. They are connected with various data collection terminals through internet networks and other communication networks. Data about an object is taken in real time; it is then converted into the appropriate data format and then sent to the data center. The data is then processed using clouds computing and other intelligent computing technologies that can process and store large amounts of data [2].

As many technologies involved in building up IoT, a reliable data communication system is needed to protect every part of the data. Broadly speaking, there are three things that may become an issue. The first is physical security, especially the security of sensors and RFID from interference, and a signal interception. Second is that all of those elements such as sensors, transmission systems, and others, should operate normally. The system should guarantee that all those elements should work as they should be. The last one is reliable data and information exchange. The message transmission between devices in IoT should be more reliable, authentic, has integrity and confidentiality. The data and information sent and received along the internet network should not be damaged, stolen or falsified [2].

Because of this, the reliability of the IoT systems can be enhanced by adding an error correction scheme both in the communication channel as well as the data store. By introducing the error correction scheme, the risks may be reduced to an acceptable level, and the reliability could be enhanced. Reed-Solomon (RS) code is one of many error control coding schemes available. It was firstly introduced by Reed and Solomon in 1960. This code has been used in various applications, such as CD-ROMs, space communications, DVD technology, digital TV and much more.

2. Error Control Codes

The general idea behind error control coding is to put some restrictions on the characteristic of the source signals so that the transmitted signals are more immune to corruption. Therefore the effects of noise during transmission are reduced or even eliminated. The term coding refers to error detection and error correction techniques. Hence, the work of error control coding involves a technique by which some extra information is added to the message signal in such a way that allows the receiver to estimate the transmitted data from the corrupted received information. This extra information is used to shield the message signals.

The work of coding was pioneered by Shannon in 1948 [3]. He showed that the existence of error control codes would reduce the existence of data corruption in the channel if data is transmitted at a rate less than the channel capacity. Reed-Solomon (RS) codes, developed by Irving Reed and Gus Solomon in 1960 [4], are a type of error control code which have proven to be of practical value in a variety of applications ranging from satellite communications to the recording of information on a storage device [6, 7, 8, 9, 10].

A typical data communication system is shown in Figure 1. The RS encoder takes a block of digital data and then adds extra 'redundant' bits. Errors may occur during transmission or storage for a number of reasons, for example, cross talk on a telephone line scratches on a CD, etc. The RS decoder processes each block and attempts to correct errors and recover the original data. The number and type of errors that can be corrected depending on the characteristics of the RS code.

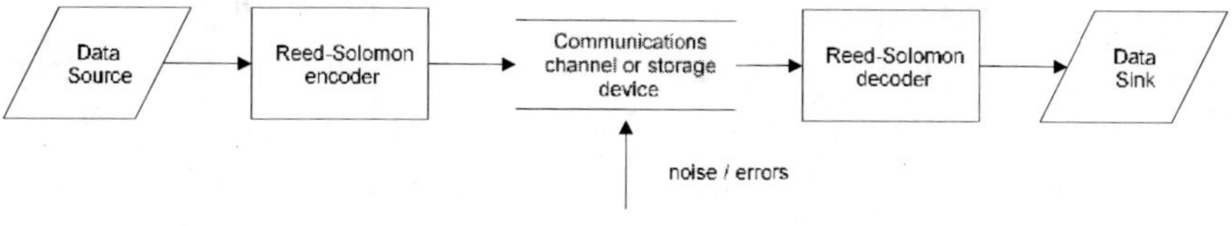

Figure 1. The Digital Communication system.

The RS code is specified as RS(n, k) over GF(2^m) with m bits per symbol. This means that

n (= $2m$-1) is the number of symbols in every codeword
k (= $n - 2t$) is the number of information symbols in every codeword

If t is the maximum number of symbols that can be corrected

d (= $2t + 1$) is the designed distance
$2t$ (= n-k) is the number of parity check symbols

then the RS code is shown in Figure 2.

Data = k symbols	Parity = n-k symbols

Figure 2. The systematic RS code.

The RS encoder works by taking k data symbols of m bits each and adds n-k parity symbols to make an n symbol codeword. If the information symbol is left unaltered and the parity symbols are appended (as shown in Figure 2), it is called a systematic encoder. It is different from a non-systematic encoder where the information symbols have been altered in such a way that the information symbols are not obvious [11, 12]. This non-systematic encoder is not covered in this chapter. The amount of processing required to encode and decode RS codes depends on the number of symbols that can be corrected. A large value of t means that a large number of errors can be corrected, but it requires more computation and more hardware than a small value of t.

3. RS Encoder

The RS(n, k) code of block length n, message length k, and designed distance n-k+1, is easily explained by using its special polynomial, called generator polynomial, $g(x)$. The general form of the RS(n, k) generator polynomial is shown in equation (1) [13, 14, 15].

$$g(x) = \prod_{i=0}^{n-k-1} (x - \alpha^{b+i}) \tag{1}$$

where b is a nonnegative integer selected by the designer, without loss of generality, it is assumed that $b = 0$.

All valid RS(n, k) codewords can be regarded as being polynomials of degree (n-1) over GF(2^m) which are divisible by its generator polynomial g(x). Let

$$I(x) = i_0 + i_1 x + ... + i_{k-1} x^{k-1} 7 \tag{1}$$

be the information polynomial to be encoded. The systematic encoding can be considered as multiplying the information polynomial $I(x)$ by x^{n-k} and then follows by dividing the result with $g(x)$ to obtain the remainder polynomial of degree n-k-1. The encoded signal is then formed by subtracting $r(x)$ from $I(x)x^{n-k}$. It can be written as equation (2) [12, 15, 16]

$$c(x) = I(x).x^{n-k} + (I(x).x^{n-k} \bmod g(x)) \tag{2}$$

It should be noted that the remainder polynomial is, in fact, the parity check polynomial. This process ensures that a valid codeword $c(x)$ is a multiple of $g(x)$. The information symbol appears explicitly at the k highest order coefficients and the parity check symbols at the n-k lowest order coefficients. From this fact, it can be seen that the process of encoding is basically carrying out division over a finite field GF(2^m).

For example, if $g(x) = 1 + x + x^3$ and $I(x) = b_0 + b_1 x + b_2 x^2$. Multiplying $I(x)$ by x^3 will result $b_0 x^3 + b_1 x^4 + b_2 x^5$. If this result is divided by $g(x)$, then a quotient $(b_0 + b_2) + b_1 x + b_2 x^2$ and a remainder $(b_0 + b_2) + (b_0 + b_1 + b_2)x + (b_1 + b_2)x^2$. A circuit, which implements this computation, is shown in Figure 3 [4, 12].

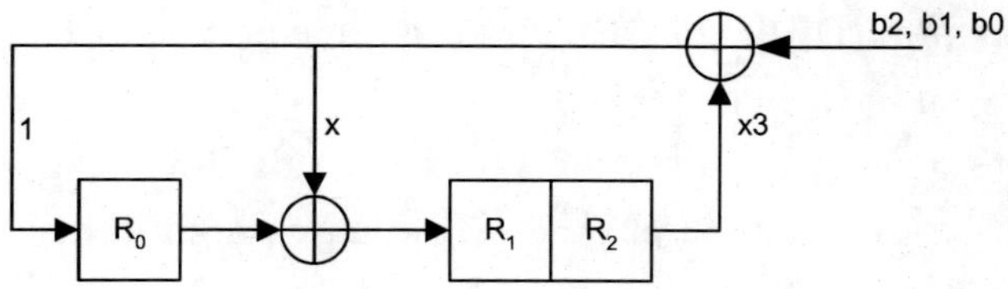

Figure 3. The structure of RS encoder over GF(2).

Shift 0).	$R_0 = 0$	$R_1 = 0$	$R_2 = 0$
Shift 1).	$R_0 = b_2$	$R_1 = b_2$	$R_2 = 0$
Shift 2).	$R_0 = b_1$	$R_1 = b_1 + b_2$	$R_2 = b_2$
Shift 3).	$R_0 = b_0 + b_2$	$R_1 = b_0 + b_1 + b_2$	$R_2 = b_1 + b_2$

The circuit in Figure 3 is very simple because it operates over GF(2) only. In general case of a code defined over GF(2^m), multiplication over GF(2^m) must also be performed. This generalized encoding scheme is shown in Figure 4 [13, 15].

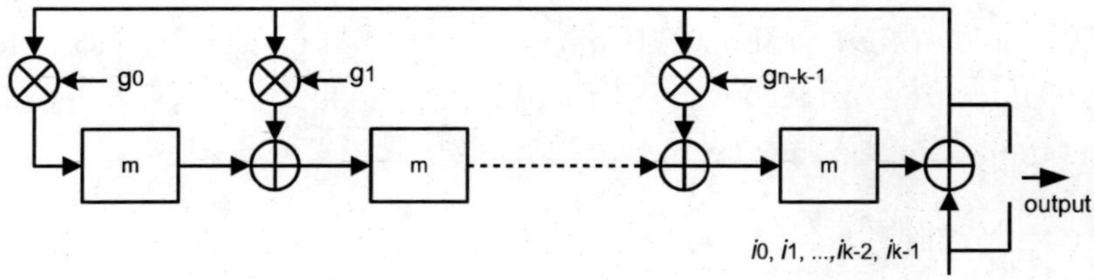

Figure 4. The structure of RS encoder over GF(2^m).

4. RS DECODER

In this section, the decoding method of RS codes is considered. The decoding process can be generally classified into two categories, hard decision decoding (HDD) algorithm, and soft decision decoding (SDD) algorithm. HDD algorithms include frequency domain decoding and time domain decoding [12, 18, 19]. Frequency domain decoding begins with the calculation of the syndromes which is the finite field equivalent of a Discrete

Fourier Transform (DFT) [17]. However, in time domain decoding there is no calculation of the syndromes.

There are quite a few algorithms to decode the RS codes, such as the Euclidean algorithm [20, 21, 22], the Fitzpatrick algorithm [23, 24], the Berlekamp Massey (BM) algorithm [25] and the Welch Berlekamp (WB) algorithm [26]. Here, the WB algorithm developed by L. Welch and E. R. Berlekamp is considered. Many researchers have commented and investigated this algorithm, as can be seen in [22, 27, 28, 29, 30, 31, 32]. This decoding method does not involve syndrome computation as a starting point. As in [30], this scheme may be used to develop the GMD decoder, one of the many approaches in applying soft decision decoding.

5. RS Decoding Using the WB Algorithm

The WB algorithm performs RS decoding without explicit syndrome computation. Furthermore, the WB algorithm is known to be more amenable to SDD which offers around 2dB more coding gain than HDD [13]. Here, the WB algorithm for decoding RS code is presented [26, 30]. The algorithm of decoding RS codes can be broken down into 4 stages as shown in Figure 5:

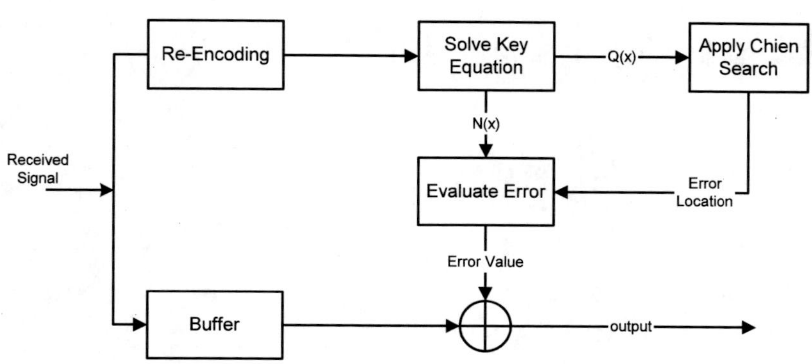

Figure 5. The RS decoder using the WB algorithm.

Stage 1: Re-encoding.
Stage 2: Solving the key equation using the Welch-Berlekamp Algorithm.
Stage 3: Applying the Chien search to solve the error locator polynomial.
Stage 4: Evaluating the error value to obtain the corrected message.

The first stage of decoding is re-encoding the received input signal $v(x)$. It corresponds to the syndrome computation in conventional RS decoding[29]. Basically, this process is quite similar to the encoding process. Re-encoding takes received information signals and then internally generates $2t$ parity check signals. If $r'(x)$ represents the internally generated parity check signals, then the re-encoding output will be the mod 2 addition of r'_i and the parity check input, v_i. This re-encoding output is called remainder polynomial $r(x)$. It can be expressed as equation (3) and (4), and the re-encoder circuitry with generator polynomial $g(x)$ as in (1) is shown in Figure 6.

$$r_i = \begin{cases} 0 : n-1 \geq i \geq 2t \\ v_i + r_i^{'} : i < 2t \end{cases} \quad 2828 \tag{3}$$

$$r(x) = \sum_{i=0}^{2t-1} r_i x^i \ 2929 \tag{4}$$

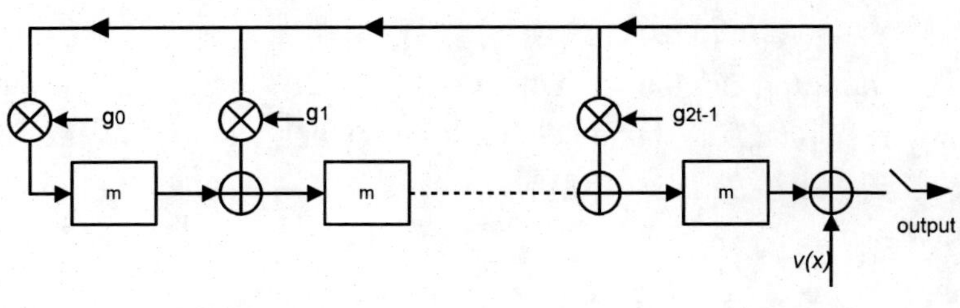

Figure 6. The re-encoder circuitry.

Let

$$g'(x) = \prod_{i=0}^{2t-2} (x - \alpha^{i})\ 3030 \tag{5}$$

and then equation (6) is pre-computed constants held in memory.

$$G(x) = \sum_{i=0}^{2t-1} G_i x^i = \sum_{i=0}^{2t-1} \frac{1}{g_i^{'}} x^i\ 31 \tag{6}$$

Then, R_i is computed to get the Welch-Berlekamp input [29] as expressed in equation (7).

$$R_i = r_i G_i\ 32 \tag{7}$$

for $i = 0, 1, \ldots, 2t$-1.

Then, the Welch Berlekamp algorithm is used to solve the key equation. Let the input to WB algorithm be $2t$ pairs of (R_k, α_k) where R_k corresponds to the parity check input and α_k $(=\alpha^k$, for the case of error, only decoding) corresponds to the check location. As in [25, 26, 29], the WB algorithm is presented in this section.

For two polynomials $Q(x)$ and $N(x)$ define $L[Q(x), N(x)]$ as the length of the pair, which is the higher value between the degree of $Q(x)$ and degree of $N(x) + 1$ [26]. A similar approach applies to both $W(x)$ and $V(x)$ polynomials and its corresponding $L[W(x), V(x)]$.

At the iteration number p (where $0 < p < 2t$), there are two pairs of polynomials $[Q_p(x), N_p(x)]$ and $[W_p(x), V_p(x)]$ which already fits the already processed checks in equations (8) and (9).

$$Q^{p}(\alpha_{p-1}) R_{p-1} = N^{p}(\alpha_{p-1})\ 33 \tag{8}$$

$$W^{p}(\alpha_{p-1}) R_{p-1} = V^{p}(\alpha_{p-1})\ 34 \tag{9}$$

Furthermore, $L[Q(x), N(x)]$ is less than $L[W(x), V(x)]$. The algorithm needs to be extended from p to $p+1$.

At first, the pair of polynomials $[Q^p(x), N^p(x)]$ needs to be checked if it can accommodate the new input pair. The way to do it is by computing D1 as shown in equation (10)

$$D1 = Q^p(\alpha_{p+1})R_{p+1} - N^p(\alpha_{p+1}) 35 \tag{10}$$

If $D1 = 0$, it means that $[Q^p(x), N^p(x)]$ can accommodate the new input pair (R_{p+1}, α_{p+1}) and then branch A of Figure 7 is taken. However, if $D1$ is not zero, then $D2$ needs to be computed as shown in equation (11).

$$D2 = W^p(\alpha_{p+1})R_{p+1} - V^p(\alpha_{p+1}) 36 \tag{11}$$

Then, the two pairs $[Q^p(x), N^p(x)]$ and $[W^p(x), V^p(x)]$ need to be updated. The first updated pair is the $(W, V)^{p+1}$ polynomials as shown in equation (12)

$$(W,V)^{p+1} = (W,V)^p + D2(Q,N)^p / D1\, 37 \tag{12}$$

and the second pair is the $(Q, N)^{p+1}$ polynomials as shown in equation (13)

$$(Q,N)^{p+1} = (Q,N)^p (x - \alpha_{p+1}) 38 \tag{13}$$

Finally, the orders of the two updated pairs need to be compared. The shorter value between equations (12) and (13) is set to be $[Q^{p+1}(x), N^{p+1}(x)]$, and the longer one is set to be $[W^{p+1}(x), V^{p+1}(x)]$. In other words, if the shorter polynomial pair is given by equation (13), then branch B of Figure 7 is taken. However, if the shorter polynomial pair is given by equation (12), then branch C is taken.

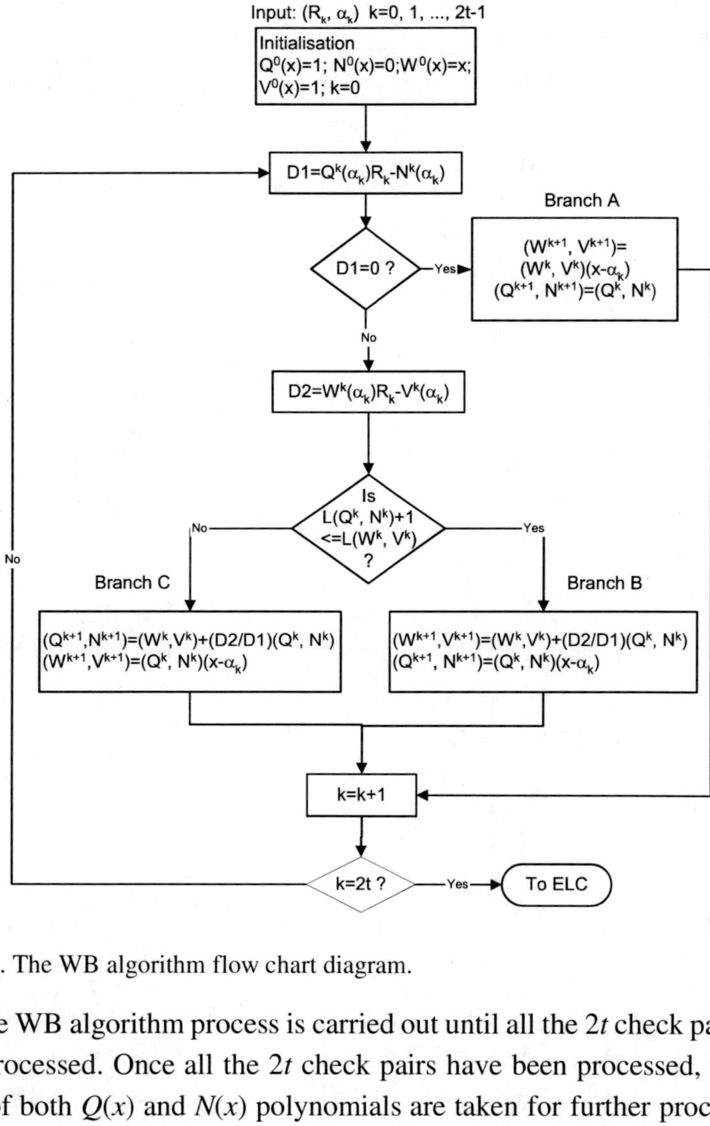

Figure 7. The WB algorithm flow chart diagram.

The WB algorithm process is carried out until all the $2t$ check pairs have been processed. Once all the $2t$ check pairs have been processed, the final value of both $Q(x)$ and $N(x)$ polynomials are taken for further process. The $Q(x)$ polynomial represents the error locator polynomial, and its roots need to be found. The Chien search method [32] is used to solve the $Q(x)$ polynomial. Let $Q(x)$ be a polynomial of degree p (where $p \leq t$) as shown in equation (14).

$$Q(x) = q_p x^p + q_{p-1} x^{p-1} + \ldots + q_0 \, 39 \qquad (14)$$

If $Q_l(x)=q_l x^l$ for $0<l\leq p$, then equation (15) applies

$$Q(x)=Q_p(x)+Q_{p-1}(x)+...+q_0 40 \tag{15}$$

and equation (15) can be shown as equation (16).

$$Q(\alpha^x)=Q_p(\alpha^{x+1})\alpha^{-p}+Q_{p-1}(\alpha^{x+1})\alpha^{1-p}+...+q_0 41 \tag{16}$$

The implementation of equation (16) above requires p multipliers. As the $Q^{2t}(x)$ polynomial will be of the order up to t, it requires t multipliers.

In order to find the error value, more pre-computed constants are required [29]. Let

$$C=[(\alpha^0\alpha^1...\alpha^{2t-2})(\alpha^0-\alpha^1)...(\alpha^0-\alpha^{2t-1})]^{-1} 42 \tag{17}$$

and the C value is pre-computed. Then, the C value is multiplied by the generator polynomial $g(x)$ as shown in equation (18).

$$f(x)=Cg(x) 43 \tag{18}$$

The $f(x)$ polynomial is a pre-computed constant held in memory. This $f(x)$ value is solved using a method shown in (16). It requires $2t$ multipliers.

Once the WB algorithm has been completed, it is followed by the error correction and error evaluation. There are two possible error positions. The first case is error occurred at a data location and its value is obtained using equation (19) below. Where $Q'(x)$ polynomial is the derivative of $Q(x)$ polynomial.

$$e_i=\frac{N(\alpha^i)}{f(\alpha^i)Q'(\alpha^i)} 44 \tag{19}$$

The second case is error occurred at a check location. The error value at the check location can be found using equation (20) below.

$$e_i = \frac{N(\alpha^i)}{f(\alpha^i)Q'(\alpha^i)} + v_i \text{, where } v_i \text{ is the received data.4} \qquad (20)$$

Alternatively, it could be corrected by re-encoding the error-free information signal or not to be corrected at all.

6. RS Modified WB Algorithm

Consider the RS(n, k) decoder with minimum distance d, error correcting capabilities t operating over GF(2^m). Let the input to the WB algorithm be 2t check pairs (R_k, α_k) where R_k and α_k represent the check value and check location respectively. The modified version of the WB algorithm is shown in Figure 8, which is a modified version of the one given in [30, 34]. This modification will be discussed in detail.

The most obvious difference is the use of J control signal instead of $L[Q(x), N(x)]$ and $L[W(x), V(x)]$. Let $L[Q(x), N(x)]$ is defined to be the higher value between the order of $Q(x)$ polynomial and the order of $N(x) + 1$ polynomial. Similarly applies to $L[W(x), V(x)]$, which is defined as the higher value between the order $W(x)$ and order $V(x) + 1$. The reason is explained as follows.

Let

$$J^k = L[W^k(x), V^k(x)] - L[Q^k(x), N^k(x)] \, 59{,}1 \qquad (21)$$

The cycle enters branch A if $D1 = 0$ (see Figure 7). At this branch, both $W(x)$ and $V(x)$ polynomials are multiplied by (x-α_k), on the other hand, the

$Q(x)$, and $N(x)$ polynomials remain unchanged. Accordingly, the degrees of both updated $W(x)$ and $V(x)$ polynomials are incremented by one; however, the orders of the updated $Q(x)$ and $N(x)$ polynomials remain unchanged. It is then expressed in mathematical terms as (22) and (23).

$$L[W^{k+1}(x), V^{k+1}(x)] = L[W^{k}(x), V^{k}(x)] + 1 \tag{22}$$

$$L[Q^{k+1}(x), N^{k+1}(x)] = L[Q^{k}(x), N^{k}(x)] \tag{23}$$

So applying equation (21) to both equations (22) and (23) yields equation (24).

$$J^{k+1} = J^{k} + 1 \tag{24}$$

Branch *B* is taken if $D1 \neq 0$ and $L[Q^k(x), N^k(x)] + 1 \leq L[W^k(x), V^k(x)]$ (see Figure 7). In this branch, both the Q(x) and $N(x)$ polynomials are multiplied by $(x\text{-}\alpha_k)$, therefore the updated degree of $Q(x)$ (and $N(x)$, if applicable) would be incremented by one. On the other hand, the $W(x)$ and $V(x)$ polynomials are added with $(D2/D1).Q(x)$ and $(D2/D1).N(x)$, consecutively. The order of both ($W(x)$, $V(x)$) polynomials does not change at all because of the condition $L[Q^k(x), N^k(x)] + 1 \leq L[W^k(x), V^k(x)]$. In mathematical terms, it can be expressed as equations (25) and (26)

$$L[W^{k+1}(x), V^{k+1}(x)] = L[W^{k}(x), V^{k}(x)] \tag{25}$$

$$L[Q^{k+1}(x), N^{k+1}(x)] = L[Q^{k}(x), N^{k}(x)] + 1 \tag{26}$$

Then, applying equation (21) to both equations (25) and (26) yields (27)

$$J^{k+1} = J^{k} - 1 \tag{27}$$

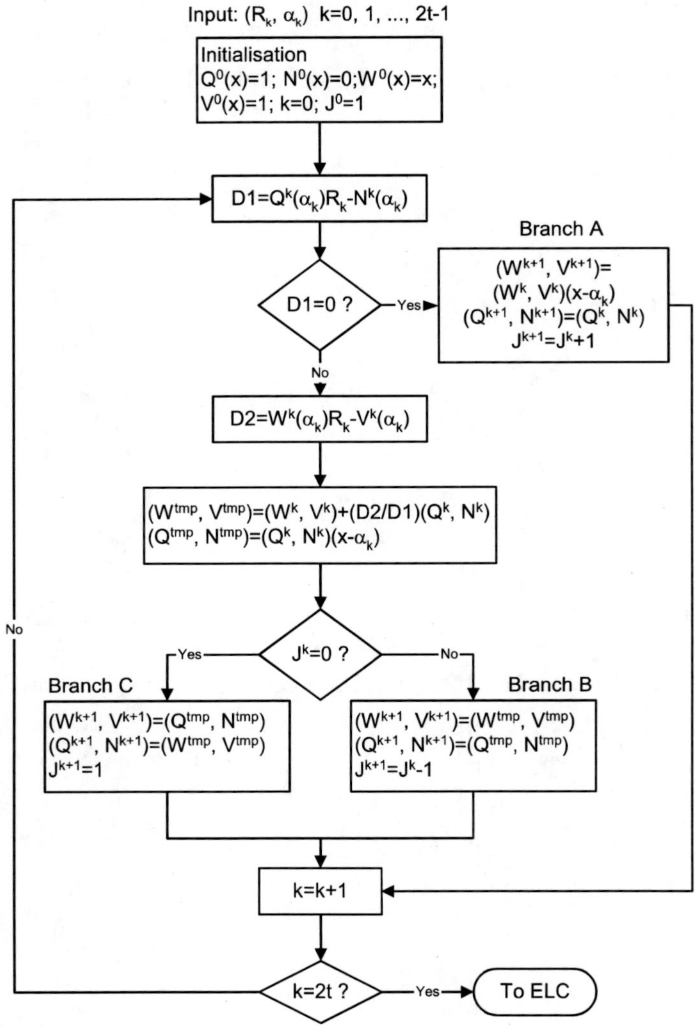

Figure 8. The modified WB algorithm flow chart diagram.

If $D1 \neq 0$ and $L[Q^k(x), N^k(x)] + 1 > L[W^k(x), V^k(x)]$, the process enters branch C (see Figure 7). The condition as in (28) can be rewritten as equation (29)

$$L[Q^k(x), N^k(x)] + 1 > L[W^k(x), V^k(x)] \tag{28}$$

$$L[W^k(x),V^k(x)]-L[Q^k(x),N^k(x)]<1 \tag{29}$$

As $L[Q^k(x), N^k(x)]$ is never more than $L[W^k(x), V^k(x)]$ at any value of k, then equation (29) could be written as (30) or (31)

$$L[W^k(x),V^k(x)]-L[Q^k(x),N^k(x)]=0 \tag{30}$$

$$J^k=0 \tag{31}$$

Then, the control signal to enter either branch B or branch C is modified to checking whether $J^k = 0$ or not. The process at branch C generates equation (32), which can be written as equation (33), and equation (34).

$$L[Q^{k+1}(x),N^{k+1}(x)]=L[W^k(x),V^k(x)] \tag{32}$$

$$L[Q^{k+1}(x),N^{k+1}(x)]=L[Q^k(x),N^k(x)], \tag{33}$$

$$L[W^{k+1}(x),V^{k+1}(x)]=L[Q^k(x),N^k(x)]+1. \tag{34}$$

Considering equation (21) to both equations (33) and (34), then it generates equation (35).

$$J^{k+1}=J^k+1. \tag{35}$$

It can be written as (36)

$$J^{k+1}=1 \tag{36}$$

The other improvement is the calculation of the intermediary stage (W^{tmp}, V^{tmp}) and (Q^{tmp}, N^{tmp}) if $D1 \neq 0$ as in Figure 8. This calculation may save hardware requirements to implement the WB algorithm and also to make the implementation easier.

7. The Implementation of the WB Algorithm

Assume that the WB algorithm operates over GF(2^8). The input check pair is (R_k, α_k) where R_k corresponds to the parity check value, and α_k corresponds to the check location. In the case of HDD, $\alpha_k = \alpha^k$. Here, four clock cycles are required to process every check pair. The first clock cycle is to compute the *D*1 value, the second cycle is to compute both *D*2 and *D*2/*D*1 values, the third one is to compute the (Q^{tmp}(x), $N^{tmp}(x)$) polynomials and finally, the fourth cycle is to calculate the ($W^{tmp}(x)$, $V^{tmp}(x)$) polynomials and to decide which pair become the $Q^{k+1}(x)$ and $N^{k+1}(x)$ polynomials and which pair become $W^{k+1}(x)$ and $V^{k+1}(x)$ polynomials.

7.1. D1 Calculation

As the error correcting capability of the decoder is 8, the maximum degree of $Q(x)$ and $N(x)$ polynomials are 8 and 7 respectively. Assume that α_k^2, α_k^3, …, α_k^8 are already available (see section 6.5 later). To obtain the $Q^k(\alpha_k)$ value, 8 Berlekamp multipliers are required as (37).

$$Q^k(\alpha_k) = Q_8^k.(\alpha_k^8) + Q_7^k.(\alpha_k^7) + Q_6^k.(\alpha_k^6) + \ldots + Q_0^k\ 60{,}2 \qquad (37)$$

Similarly, the implementation of $N^k(\alpha_k)$ calculation would require 7 multipliers as shown in (38).

$$N^k(\alpha_k) = N_7^k.(\alpha_k^7) + N_6^k.(\alpha_k^6) + N_5^k.(\alpha_k^5) + \ldots + N_0^k\ 60{,}2 \qquad (38)$$

Once both $Q^k(\alpha_k)$ and $N^k(\alpha_k)$ values are available, *D*1 can be easily calculated by implementing equation (39)

$$D1 = Q^k(\alpha_k)R_k - N^k(\alpha_k)\ 35 \qquad (39)$$

and 1/*D*1 can be easily obtained.

7.2. D2 Calculation

As with the *D*1 calculation, the error correcting capabilities of the decoder is 8. Hence, the maximum degree of both *W*(*x*) and *V*(*x*) polynomials are 8 and 7, respectively. It is assumed that α_k^2, α_k^3, …, α_k^8 are already available (see section 6.5 later). The multipliers used in *D*1 calculation in section 6.1 are re-used in this stage to implement equation (40).

$$W^k(\alpha_k) = W_8^k.(\alpha_k^8) + W_7^k.(\alpha_k^7) + W_6^k.(\alpha_k^6) + \ldots + W_0^k \quad (40)$$

Therefore 8 Berlekamp multipliers are required to compute the $W^k(\alpha_k)$ value. Similarly, the implementation of $V^k(\alpha_k)$ calculations would require 7 multipliers as in (41).

$$V^k(\alpha_k) = V_7^k.(\alpha_k^7) + V_6^k.(\alpha_k^6) + V_5^k.(\alpha_k^5) + \ldots + V_0^k \, 60{,}2 \quad (41)$$

Once both $W^k(\alpha_k)$ and $V^k(\alpha_k)$ values are available, *D*2 can be found by using (42) and then D2/D1 can be calculated.

$$D2 = W^k(\alpha_k)R_k - V^k(\alpha_k)\,35 \quad (42)$$

7.3. D1 Calculation of (($W^{k+1}(x)$ & $V^{k+1}(x)$) or ($Q^{tmp}(x)$ and $N^{tmp}(x)$))

If *D*1 = 0, then equations (43) and (44) are computed.

$$W^{k+1}(x) = \alpha_k(W_8^k x^8 + W_7^k x^7 + \ldots + W_0^k) + (W_7^k x^8 + W_6^k x^7 + \ldots + 0) \quad (43)$$

$$V^{k+1}(x) = \alpha_k (V_7^k x^7 + V_6^k x^6 + \ldots + V_0^k) + (V_6^k x^7 + V_5^k x^6 + \ldots + 0) \quad (44)$$

The numbers of multipliers required to calculate ($W^{k+1}(x)$, $V^{k+1}(x)$) polynomials are 9 and 8; respectively. Here, the 8 multipliers to calculate $W^k(\alpha_k)$ in section 6.2 are re-used to implement the $W^{k+1}(x)$ circuitry with one extra multiplier is added. Similarly, the multipliers to calculate $V^k(\alpha_k)$ in section 6.2 are also re-used to implement the $V^{k+1}(x)$ circuitry with one extra multiplier is added. There is no required update on ($Q^k(x)$, $N^k(x)$) polynomials, hence ($Q^{k+1}(x)$, $N^{k+1}(x)$) polynomials are already available. J^{k+1} value needs to be updated, hence $J^{k+1} = J^k + 1$.

On the other hand, when $D1 \neq 0$, then the following equations will be computed and the hardware used will be the one on the case of $D1 = 0$.

$$Q^{tmp}(x) = \alpha_k (Q_8^k x^8 + Q_7^k x^7 + \ldots + Q_0^k) + (Q_7^k x^8 + Q_6^k x^7 + \ldots + 0) \quad (45)$$

$$N^{tmp}(x) = \alpha_k (N_7^k x^7 + N_6^k x^6 + \ldots + N_0^k) + (N_6^k x^7 + N_5^k x^6 + \ldots + 0) \quad (46)$$

7.4. Calculation of $W^{tmp}(x)$ and $V^{tmp}(x)$

At this stage, the $W^{tmp}(x)$ and $V^{tmp}(x)$ polynomials have to be found. The way to get the $W^{tmp}(x)$ polynomial is by multiplying $D2/D1$ value with $Q^k(x)$ polynomial, and then adding the $W^k(x)$ polynomial, as shown in equation (47).

$$W^{tmp}(x) = \frac{D2}{D1}(Q_8^k x^8 + Q_7^k x^7 + \ldots + Q_0^k) + (W_8^k x^8 + W_7^k x^7 + \ldots + W_0^k) \quad (47)$$

The number of multipliers required is 9, and it will re-use the multipliers to implement $W^{k+1}(x)$ or $Q^{tmp}(x)$ in section 6.3. Similar fashion is applicable to find $V^{tmp}(x)$ polynomial using equation (48).

$$V^{tmp}(x) = \frac{D2}{D1}(V_7^k x^7 + V_6^k x^6 + \ldots + V_0^k) + (N_7^k x^7 + N_6^k x^6 + \ldots + N_0^k) \tag{48}$$

The number of multipliers required is 8 which have already been used to implement $V^{k+1}(x)$ or $N^{tmp}(x)$ on section **c**. By considering the J^k value, the $W^{k+1}(x)$, $Q^{k+1}(x)$, $V^{k+1}(x)$ and $N^{k+1}(x)$ polynomials are then updated.

If $J^k = 0$ then equation (49) applies, otherwise equation (50).

$$\begin{aligned} \left(W^{k+1}(x), V^{k+1}(x)\right) &= \left(Q^{tmp}(x), N^{tmp}(x)\right) \\ \left(Q^{k+1}(x), N^{k+1}(x)\right) &= \left(W^{tmp}(x), V^{tmp}(x)\right) \end{aligned} \tag{49}$$

and

$$\begin{aligned} \left(Q^{k+1}(x), N^{k+1}(x)\right) &= \left(Q^{tmp}(x), N^{tmp}(x)\right) \\ \left(W^{k+1}(x), V^{k+1}(x)\right) &= \left(W^{tmp}(x), V^{tmp}(x)\right) \end{aligned} \tag{50}$$

7.5. Finite Field Exponentiation

In section 6.1 and 6.2, it has been assumed that $\alpha_k^2, \alpha_k^3, \ldots, \alpha_k^8$ are available. In this section, it will be explained how to obtain $\alpha_k^2, \alpha_k^3, \ldots, \alpha_k^8$ by performing multiplication less than 7 times.

Consider $GF(2^8)$ with $p(x) = x^8 + x^4 + x^3 + x^2 + 1$ and take α to be a root of $p(x)$. Let

$$x = \begin{pmatrix} x_0 \\ x_1 \\ x_2 \\ x_3 \\ x_4 \\ x_5 \\ x_6 \\ x_7 \end{pmatrix}, \text{ then it can be shown that } x^2 = \begin{pmatrix} x_0 + x_4 + x_6 + x_7 \\ x_7 \\ x_1 + x_4 + x_5 + x_6 \\ x_4 + x_6 \\ x_2 + x_4 + x_5 + x_7 \\ x_5 \\ x_3 + x_5 + x_6 \\ x_6 \end{pmatrix}$$

$$x^4 = \begin{pmatrix} x_0 + x_2 + x_3 + x_6 \\ x_6 \\ x_2 + x_3 + x_4 + x_5 + x_6 \\ x_2 + x_3 + x_4 + x_6 + x_7 \\ x_1 + x_2 + x_5 + x_7 \\ x_5 \\ x_3 + x_4 \\ x_3 + x_5 + x_6 \end{pmatrix} \text{ and } x^8 = \begin{pmatrix} x_0 + x_1 + x_3 + x_4 + x_7 \\ x_3 + x_5 + x_6 \\ x_1 + x_2 + x_3 + x_4 + x_6 + x_7 \\ x_1 + x_2 + x_3 + x_4 + x_5 + x_7 \\ x_1 + x_4 + x_7 \\ x_5 \\ x_2 + x_5 + x_6 + x_7 \\ x_3 + x_4 \end{pmatrix}$$

The combination of this method and multiplication is carried out to perform $x^{k'}$ (where $k' \neq 2, 4, 8$).

8. VHDL Decoder Using the WB Algorithm

8.1. Pin Out Connection

The HDD pins both input and output connections are as follows.

1. parity_number : integer input 8.
2. Input *clk*: clock active on the rising edge.
3. Input *reset_n*: active low asynchronous reset.

4. Input *reset_sync_n*: active low synchronous reset.
5. Input *di*: 8-bits input data to be decoded.
6. Input *di_rdy*: active high flag to indicate that *di* is valid.
7. Input *di_sync_in*: active high flag to synchronize the codeword.
8. Input *do_acpt*: active high flag to indicate that *do* is already taken.
9. Output *do*: the 8 bits decoder output.
10. Output *do_rdy*: active high flag to indicate that *do* is valid.
11. Output *di_acpt*: active high flag to indicate that *di* is already taken.
12. Output *do_sync_out*: active high flag to synchronize the output.
13. Output *do_error*: active high flag when a correction took place.
14. Output *do_parity_n*: active low flag at a parity check location.

8.2. Timing

The decoder timing diagram is shown in Figure 9. The first stage of the decoding is re-encoding which occupies 1 codeword period. Solving the key equation is carried out at the next stage using the WB algorithm. The allocated time is also 1 codeword period, even though the WB algorithm process only requires $8t$ clock cycles. There are $2t$ check pairs need to be solved, and every pair requires 4 clock cycles to proceed. After completing the algorithm process, the Chien search is carried out to solve the error locator polynomial and then the error value is computed to recover the corrupted message.

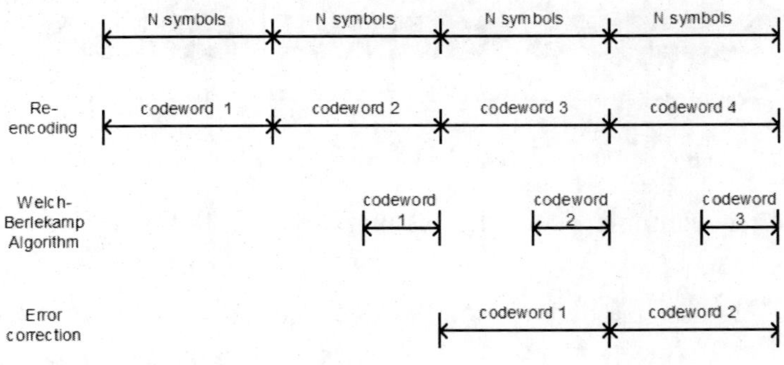

Figure 9. The timing diagram of *the RS* Decoder using the WB algorithm.

8.3. Simulation Results

Consider RS(255, 239). Let $c(x)$ be the transmitted message, and $v(x)$ be the received message.

$$c(x) = \{0, 0, 0, 0, 0, 0, 0, 0, 0, 0, 0, 0, \ldots, 0\}$$
$$v(x) = \{1, 2, 4, 8, 16, 32, 64, 128, 0, 0, 0, 0, \ldots, 0\}$$

The simulation result was shown in Figure 10 where all zeros became the decoder output. $v(x)$ is represented as *din_ff*, the error value is *value*, the corrected value is *corrected*, and the decoder output is *wb_do*.

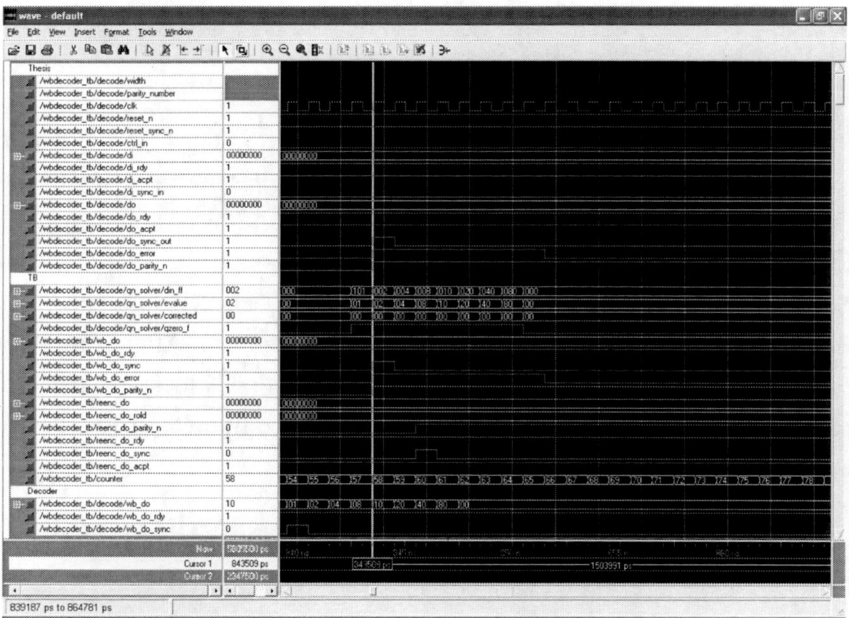

Figure 10. The simulation result for HDD RS(255, 239).

Another example, consider again RS(255, 239). Let $I(x)$ be the information message before encoding at the value of {0x0F, 0x0F, 0x0F, …, 0x0F}. Then the encoded transmitted message will be $c(x)$ = {0x0F, 0x0F, 0x0F, 0x0F, 0x0F, 0x0F, 0x0F, 0x0F, 0x0F, 0x0F, 0x0F, …, 0x0F, 0xA9, 0xCC, 0x91,

0x43, 0x2C, 0x7C, 0x63, 0x7B, 0x22, 0x55, 0x01, 0x87, 0xF1, 0x2B, 0xB1, 0x6A}.

Let $e(x)$ be the error value, and $v(x)$ be the received message, then

$e(x)$ = {0x03, 0x03, 0x03, 0x03, 0x03, 0x03, 0x03, 0x03, 0, …, 0, 0, 0, 0, 0, 0, 0, 0, 0, 0, 0, 0, 0, 0, 0, 0, 0, 0, 0}. $v(x) = c(x) + e(x)$ = {0x0C, 0x0C, 0x0C, 0x0C, 0x0C, 0x0C, 0x0C, 0x0C, 0x0F, …, 0x0F, 0xA9, 0xCC, 0x91, 0x43, 0x2C, 0x7C, 0x63, 0x7B, 0x22, 0x55, 0x01, 0x87, 0xF1, 0x2B, 0xB1, 0x6A}.

The decoder output *do* is shown in Figure 11 and this value is valid when *do_rdy* is set to logic '1'. Figure 12 shows that the decoder output *do* is equal to the encoded, transmitted message $c(x)$.

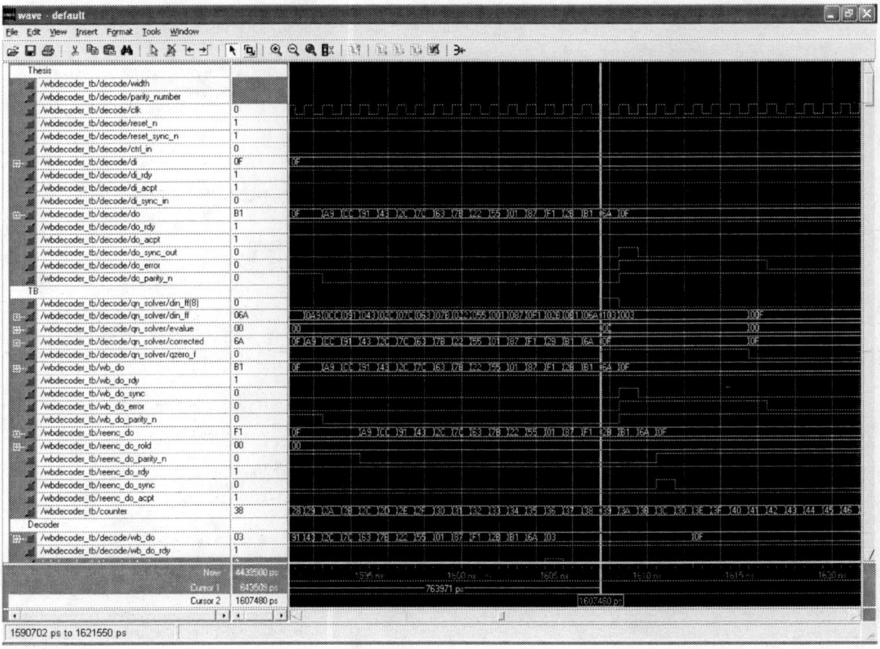

Figure 11. The simulation result for HDD RS(255, 239).

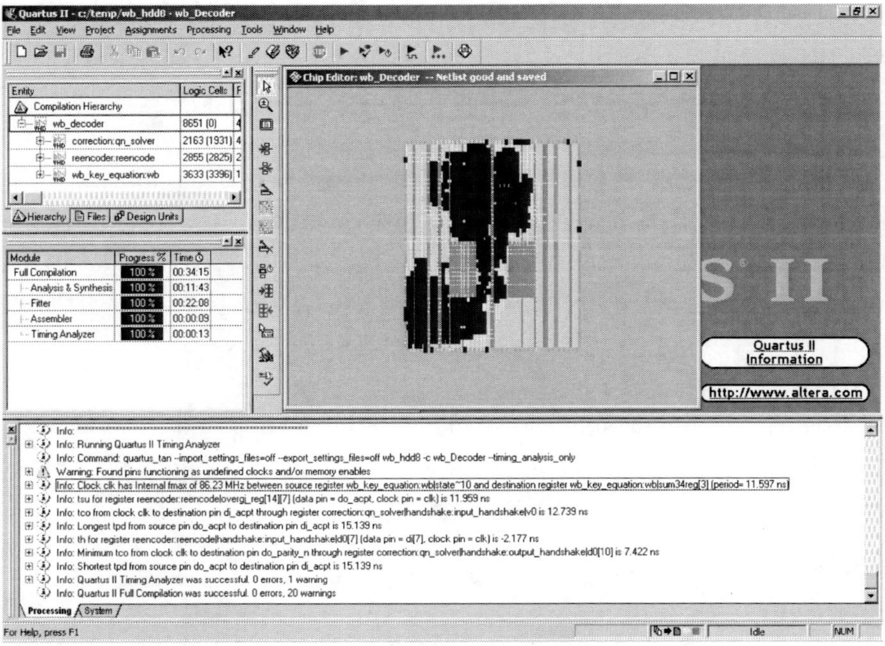

Figure 12. The synthesized result for HDD RS(255, 239).

8.4. Synthesis Results

The VHDL source code was then synthesized using *Quartus II* software. The Altera Stratix EP1S25-B672C6 was selected to be the target device with a capacity of 25660 logic element (LE). The design utilized 8651 out of 25660 (33% coverage) logic element at maximum clock frequency 86.23 MHz. A synthesize result of HDD RS(255, 239) is shown in Figure 12, where the solid black part is the utilized area.

CONCLUSION

We have presented the Reed-Solomon (RS) code as one of many error control coding schemes to enhance the security of data and information exchanges along the internet connections in the Internet of Things. Reed-

Solomon code has been discussed in details. RS decoding scheme using the Welch Berlekamp algorithm has been highlighted. The VHDL implementation of the Reed Solomon decoder is also presented. Without loss of generality, the Hard Decision Decoding for RS(255, 239) over GF(28) is developed using VHDL.

REFERENCES

[1] Meutia, Ernita Dewi. "Internet of things–keamanan dan privasi." *Semin. Nas. dan Expo Tek. Elektro* 2015 (2015): 85-89.

[2] Qiang, Chen, Guang-ri Quan, Bai Yu, and Liu Yang. "Research on security issues of the internet of things." *Issues* 371 (2016).

[3] Tiwari, A. and Maurya, H. (2017). Challenges and Ongoing Researches for IoT (Internet of Things): A Review. *International Journal of Engineering Development and Research*, vol 5, issue 2, 57-60.

[4] Shannon, C. E. (1948). A Mathematical Theory of Communication. *Bell System Tech Journal,* 27: 379-423.

[5] Reed, I. S., and Solomon, G. (1960). Polynomial codes over certain finite fields. *Journal Society. Industrial Applied Maths*, 8: 300-304.

[6] Chang, H. C. and Shung, C. B. (1998). A (208, 192; 8) Reed Solomon Decoder for DVD Application. *IEEE International Conference On Communications: Conf. Record*, affiliated with Supercomm, 2: 957-960.

[7] Hoeve, H., Timmermans, J., and Vries, L. B. (1982). Error correction and concealment in the compact disc system. *Phillips Tech. Rev*, 40, 6: 166-172.

[8] Im, Y. and Kwon, O. S. (1997). An Advanced VLSI Architecture of RS Decoders for Advanced TV. *IEEE International Conference On. Communications: Towards the Knowledge Millenium ICC'97*, 3: 1346-1350.

[9] Kwon, S. and Shin, H. (1997). An Area Efficient VLSI Architecture of a Reed-Solomon Decoder/Encoder for Digital VCRs. *IEEE Transactions on Consumer Electronics*, 43-4: 1019-1027.

[10] Shayan, Y. R., Le-Ngoc, T., and Bhargava, V. K. (1990). Design of RS (16, 12) Codec for North American Advanced Train Control System. *IEEE Transactions of Vehicular Tech*, vol. 39, no 4.

[11] Blahut, R. E. (1983). *Theory and Practice of Error Control Codes.* Reading MA, Addison-Wesley.

[12] Clark, G. C. and Cain, J. B. (1981). *Error-Correction Coding for Digital Communications.* Applications of Communications Theory, Plenum Press, New York.

[13] Berlekamp, E. R. (1982). Bit Serial Reed-Solomon Encoders. *IEEE Trans. Inform. Theory*, 28: 869-874.

[14] Bossert, M. (1999). *Channel Coding for Telecommunications.* John Wiley Publisher.

[15] Hsu, I. S., Reed, I. S., Truong, T. K., Wang, K., Yeh, C. S. and Deutsch, L. J. (1984). The VLSI Implementation of a Reed-Solomon Encoder Using Berlekamp's Bit Serial Multiplier algorithm. *IEEE Trans. Comp*, vol. 33.

[16] Liu, K. Y. (1984). Architecture for VLSI Design of a Reed-Solomon Decoder. *IEEE Trans. Comp*, vol. 33.

[17] Bhargava, W. *Reed-Solomon Codes and Their Applications*. IEEE press.

[18] Blahut, R. E. (1979). Transform Technique for Error Control Codes. *IBM Journal Res. Develop.*, vol 23 no 3, 299-315.

[19] Blahut, R. E. (1984). A Universal Reed-Solomon Decoder. *IBM Journal Res. Develop.*, vol 28 no 2, pp 150-158.

[20] Shao, H. M., Truong, T. K., Deutsch, L. J., Yuen, J. H. and Reed, I. S. (1985). A VLSI Design of a Pipeline Reed-Solomon Decoder. *IEEE Trans on Computers*, Vol C34, no. 5, pp. 393-403.

[21] Brent, R. P. and Kung, H. T. (1984). Systolic VLSI Arrays for Polynomial GCD Computation. *IEEE Trans. on Computers*, vol 33 no 8, pp 731-736.

[22] Dabiri, D. and Blake, I. F. (1995). Fast Parallel algorithms for Decoding Reed-Solomon Codes Based on Remainder Polynomials. *IEEE Trans. on Inform. Theory*, vol. 41 no 4, pp. 873-885.

[23] Fitzpatrick, P. (1996). On The Scalar Rational Interpolation Problem. *Mathematics of Control, Signals, and Systems*, vol. 9 part 4, pp 352-369.

[24] Blackburn, S. R. and Chambers, W. G. (1996). Some Remarks on an algorithm of Fitzpatrick. *IEEE Trans. on Inform. Theory*, vol 42 no 4, pp 1269-1271.

[25] Massey, J. L. (1969). Shift-Register Synthesis and BCH Decoding. *IEEE Trans. Inform. Theory*, vol. 15, no. 1, pp. 122-127.

[26] Welch, L. and Berlekamp, E. R. (1986). *Error Correction for Algebraic Block Codes*. US Patent 4,633,470.

[27] Morii, M. and Kasahara, M. (1992). Generalized Key Equation of Remainder Decoding algorithm for Reed Solomon Codes. *IEEE Trans. Inform. Theory*, vol. IT 38, 1801-1807.

[28] Chambers, W. G. (1993). A solution of Welch-Berlekamp Key Equation by Euclidean algorithm. *Electronic Letters*, vol 29 no 11, pp 1031.

[29] Chambers, W. G., Peille, R. E., Tsie, K. Y. and Zein, N. (1993). Algorithm for Solving the Welch-Berlekamp Key-Equation, with a Simplified Proof. *Electronic Letters*, vol. 29 no 18, pp 1620-1621.

[30] Peille, R. E. (1994). On the Performance and Complexity of a Generalized Minimum Distance Reed-Solomon Decoding algorithm. *International Journal of Satellite Communications*, vol. 12, pp. 333-359.

[31] Jennings, S. M. (1995). Grobner Basis View of Welch-Berlekamp algorithm for Reed-Solomon Codes. *IEE Proc. Commun.*, vol. 142 no 6, pp 349-351.

[32] Sorger, U. K. (1993). A New Reed-Solomon Code Decoding algorithm Based on Newton's Interpolation. *IEEE Trans. Inform. Theory*, vol. 39, no 2, pp. 358-365.

[33] Chien, R. T. (1964), Cyclic Decoding Procedures for BCH Codes. *IEEE Trans. Inform. Theory*, vol. 10, pp. 357-363.

[34] Mahmudi, Ali, and Sentot Achmadi. "Modified Welch Berlekamp Algorithm to Decode Reed Solomon Codes." In *MATEC Web of Conferences*, vol. 164, p. 01003. EDP Sciences, 2018.

In: Advanced Security Issues …
Editor: Vijey Thayananthan
ISBN: 978-1-53615-538-9

Chapter 3

EFFICIENT DATA TRAFFIC TECHNIQUES FOR IOT BASED 5G NETWORKS

Ahmed Alshaflut[1] and Mohammed Yahya Alzahrani[2]
[1]Department of Computer Science,
King Abdulaziz University, Jeddah, KSA
[2]Department of Information Technology,
Albaha University, Albaha, KSA

ABSTRACT

Wireless Connections and volume of the traffic are increasing with different applications over 5G environments. In fact, data traffic is significant due to the increasing numbers of services and heavy users. Thus, many researchers have discussed the increasing range of user's requirements within the 5G and 5G+ networks. The primary aim of this chapter is to provide an efficient approach to handling traffic issues in multiple access environments. This depends on achieving variable delays and data priorities associated with the large volume of the traffic considered in the 5G networks. However, traffic management plays an important role in satisfying the user's needs in future wireless communications. In this chapter, the massive data traffic will be studied for enabling the best of IoT applications over a 5G network. Therefore, the multiple accessing schemas are presented for managing the overall network traffic. Despite the previously proposed models used in the IoT based 5G networks, we can employ the traffic manager based on software define multiple access (SoDeMa) for the enhancing the network performance as well as to maintain the QoS at the best rate. Although many algorithms such as load balancing, admission control, etc. have been proposed for managing traffic problems, we have considered SoDeMa with an appropriate algorithm in our proposed work.

Keywords: data traffic, 5G, IoT based 5G, SoDeMa

INTRODUCTION

All data traffic need appropriate channels for transmitting the data and delivering on time according to the priority set by the data traffic management system. Different services of private and public organizations rely on efficient technology which provides the data traffic facilities such as quick delivery. Cognitive radio network architecture enhances not only the network traffic facilities but also QoS when service providers deliver the variable data traffic and channel conditions [1].

The motivation of this chapter is to focus on managing and estimating future network traffic which is dynamically increasing from many service providers and IoT devices. Despite many data traffic techniques, managing dynamic traffic from various sources needs efficient technique which should provide not only the data management facilities but also technical requirements according to the applications. For instance, medical applications need the highest priorities of handling data traffic in most of the emergency. In order to manage and estimate the network traffic efficiently, we introduce the novel data traffic technique based on the SoDeMa which is one of the latest multiple access technologies developed from the SDN concepts and technology.

As an efficient data traffic technique, novel data traffic scheme employs the SoDeMa with traffic algorithm which will be compatible with 5G requirements and 5G+ environments. Traffic scheduling and load balancing are growing problems in many wireless applications which employ the SDN.

For instance, the hybrid routing forwarding scheme and congestion control algorithm can be used to solve these problems in the traffic of the software-defined mobile wireless network. Artificial Fish Swarm Algorithm (AFSA) is a new intelligent optimization algorithm which is suitable for solving large-scale complex optimization problems when we consider the data traffic among the billions of IoT devices in the 5G environments. Using AFSA, traffic and link capacity of crowdsourcing tasks such as services of IoT devices can be optimized intelligently [2].

According to the chapter title, we can deliver the proposed work for future traffic solutions in IoT applications since it would assure that the service is perfectly delivered among different network circumstances. Also, discussing these issues would assure better allocation of networks resources. However, this solution will not only maintain the traffic management, but it also would inspire the service providers by considering the service lifecycle over 5G networks. Thus, it will also benefit overcoming the delay issues by estimating the frequent rush times and specific user's requirements.

State of the Art

This section provides the latest techniques of data traffic which contains the varieties data with different rates and properties. When IoT devices are increasing, the volume of the traffic will be increased according to the services users invoked to perform their duties. In order to simplify the data traffic management, we deploy efficient data traffic techniques with the latest multiple schemes such as NOMA and SoDeMa.

Data Traffic in IoT

The IoT devices exchange the data for transmitting necessary information when different networks merge with IoT environment. Although 5G IoT provides new challenges and IoT solutions as given in [4], IoT based 5G network is being developed to improve the services with efficient data traffic management. Thus, it can be modified to IoT applications within future IoT based 5G networks. Thus, the promising IoT applications have to consider the available traffic managing schema to keep enhanced network resources by keeping the delay of services at the minimum rate.

The massive amount of data gathered by IoT devices, sensors, etc. need an efficient management technique to improve the communication performance in the future smart city [6]. Despite the self-organizing technique for handover processing, the authors suggested intelligent traffic management. According to [7], relay selection procedures increase the data rate of the communication in the IoT. As far as the traffic between the IoT terminals is concerned, traffic management depends on many problems such as IoT terminals face the inter-cell interference.

A Brief Description of NOMA

According to [8], basic facts of 5G and future challenges which include the solutions, opportunities and research trends of NOMA for 5G show us that high spectral efficiency and massive connectivity are achievable using appropriate multiplexing categories of NOMA [19, 20]. In multi-user applications, complexity reduction techniques have been considered using NOMA. In addition, user-pairing algorithm provided optimal transmission scheme with little throughput loss. Survey paper [10] describes NOMA for 5G and other several multiple accessing technologies based on NOMA. In these 5G key technologies, non-orthogonal waveforms modulation techniques play many important roles with the NOMA. When the appropriate technique identifies or removes the unwanted energy for the interferences from the 5G systems, the energy saving is possible. In addition, authors have introduced the concept of NOMA which is one of the techniques to increase EE. As shown in Figure 1, we can illustrate the NOMA which accommodates the users according to the random resources which are time, frequency and power.

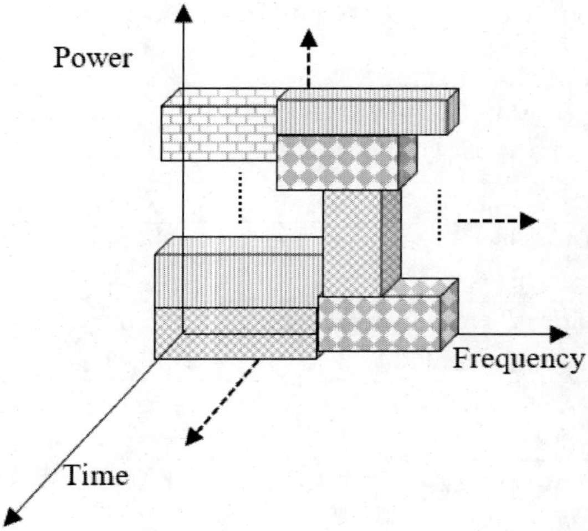

Figure 1. The NOMA with many users.

According to [3], managing communication resources with minimum latency needs a new technique. Although 5G requirements support to manage the system resources, achieving better latency is difficult because of growing communication capacity. Authors proposed a novel architecture with the cognitive engine and data engine which influence resource cognitive intelligence. In this architecture, latency can be improved through computing, caching, and communication strategies.

Different traffic models such as dynamic and static described in [5] allow us to develop a new approach which provides efficient traffic management although many challenges encouraged us to explore the traffic modeling used in IoT-femtocell based applications, some of them involved with accessing technologies such as dynamic spectrum access.

Mobile data traffic is increasing with mobile users but managing dynamic data traffic and network planning need an intelligent solution. As mentioned in [9], designing a cell-association and traffic-offloading algorithm may be the future solution for improving the traffic problems.

Table 3.1 shows the different multiple accessing methods for enhancing EE which depends on the latency. Thus, traffic can be managed efficiently with minimum latency and delay.

Table 3.1. Data traffic and 5G based multiple accessing schemes

Schemes	Requirements for SoDeMa for analyzing traffic characteristic	Overall benefits for improving data traffic
NOMA	Power allocations	High spectral efficiency
SCMA	Sparse coding schemes with non-orthogonal codes	More diversity
PDMA	Specific pattern division schemes with non-orthogonal codes	Efficient receiver with low complexity
MUMA	Legacy coding schemes with non-orthogonal codes	Better EE
BOMA	Building block sparse-constellation	Simple and low complex receiver
LPMA	Multilevel lattice code	Power and code domain

Multiple access provides a number of benefits for managing data traffic through the efficient design of multiple access protocols. They are channel partitioning, random access for brain communication, dynamic resource management, cache management, cooperative content caching, etc. Generating multiple access protocol based on SoDeMa techniques will improve not only the data traffic but also other benefits.

A Brief Description of SoDeMa

As far as heavy traffic is concerned, legacy channel models and infrastructures need efficient data traffic techniques which increase the channel capacity according to the 5G requirements. Multiple access techniques provide data handling facilities when we employ multiple types of devices with varying requirements. Although future data traffic expects to face many different priorities between the source and destination, SoDeMa split into the data traffic according to the 5G requirements (data rate, latency, out-of-band emissions, spectral efficiency, MIMO compatibility)

Definition of SoDeMa: When two or more simultaneous data traffics transmitted with different conditions share a broadcast channel, multiple accessing procedure occurs with the defined software approaches.

Intelligent SoDeMa

Decision making based on the priorities allows service providers to activate the intelligent facilities. Priorities of the data traffic vary with the data rate depending on the service of the applications such as health, transport, etc. Decision making improves the data traffic management intelligently when service providers employ the SoDeMa with decision making a feature. In this data traffic management, the role of the intelligence not only increases the speed of the data traffic but also reduces the latency.

Advances in 5G

According to [12], recent research on EE using NOMA within the 5G environment seems to corroborate with existing theories. In this research, the density function of channel gains, outage performance, and ergodic sum-rate have been used with basic theories to improve the EE. If the future 5G user can justify the claim that a given 5G system is set with ergodic sum rate, then this proposed approach based on these theories will provide an explanation for why EE is increasing.

Buzzi et al. [13] showed that EE would be achievable when wireless MIMO deployed in 5G applications. Further, MIMO was set to operate at millimeter wave frequencies (mmWave) which is one of the components to improve 5G requirements. Although 2 types of single career modems are considered, single-carrier modulation with time-domain equalization produced better performance. This approach may be used with advanced equalization based on LDPC coding developed from the parameters of the H matrix.

Despite many modulation schemes, employing a novel quadrature space-frequency index modulation in 5G wireless communication will increase not only the EE but also data rate without any extra cost of energy [14]. Further, this approach can be employed to improve the satellite channels and communication when future 5G wireless systems are deployed in the satellite modems.

According to [15], satellite links and communication need a reliable system with the best latency and less energy consumption. Here, the authors presented a simulated annealing and clustering hybrid algorithm (SACA) to enhance the reliability and to achieve the minimum latency. Regarding the design and develop the future 5G satellite modems, SACA approach will allow the satellite system designers to investigate the complexity and optimum design parameters of the 5G satellite modems.

Chen et al. [16] presented the network simulation for IMT-2020(5G) systems and satellite communication systems based on Optimized Network Engineering Tool (OPNET). Although co-channel interference from IMT Systems to satellite system is one of the design problems, reducing energy consumption and complexity needs to be investigated. Further, this network simulation will allow investigating the satellite system design issues such as satellite modulation schemes with LDPC.

Analysis of 5G LDPC codes rate-matching design created new research on energy efficient LDPC scheme for future 5G wireless communication systems which include the satellite modems [17]. Decoding latency based on the efficient design of LDPC improves not only the EE but also reliability which increases the lifetime of the 5G system. Although energy harvesting and saving are the main approaches of the future research directions, evolving technology changes keep changing the design and infrastructure of the next generation of 5G or 5G+ in the future.

As shown in Figure 2, service providers organize spectrum management with necessary emerging 5G communication technologies. Although traffic estimation and distribution are the part of the spectrum management issues, the configuration of each BS with emerging 5G communication technologies will depend on the heavy traffic.

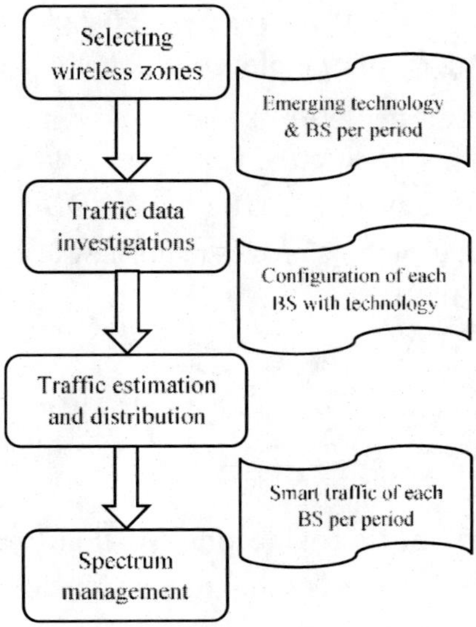

Figure 2. Flow chart of the spectrum management.

EXISTING TRAFFIC APPROACHES THROUGH SDN

Although many existing traffic models such as data traffic aggregation model, data traffic slicing model, etc. are available, this section focuses on traffic approaches using SDN. When the application layer of SDN holds the SoDeMa, management issues of the data traffic improve. Controllers are the brain of the control layer of the SDN, which organize the data traffic according to the technical properties.

Lightweight Admission Control and Traffic Management

According to [22], traffic management looks after the lightweight data which may be around 40 bytes. Lightweight admission control and traffic management with SDN allow the service providers to improve the traffic of lightweight data such as monitoring at data plane. Measurement and classification of smart systems during the data traffic over 5G mobile networks should be considered carefully with the efficient algorithms [23]. In these cases, SoDeMa may be employed to simplify and disperse the traffic according to the situation and priority.

According to [21], service providers try to reduce the cost of the data traffic while preventing unwanted traffic such as DDoS attacks. To improve the network services, the authors proposed the new approach which depends on the mobile back haul SDN.

Traffic Engineering in SDN

According to [11], traffic engineering solutions with the SDN provide efficient traffic management according to the situations. Instead of using a fixed weight setting of routing algorithms, dynamic weight setting allows the service providers to improve the traffic performance as well as network management.

According to the [18], traffic measurements considered in the network applications are prerequisite of traffic management. Despite many traffic models, traffic measurements are dynamically monitored using SDN concept which not only enhances the monitoring facilities but also manages and controls the traffic systematically.

Proposed Model for Managing Traffic

Through the literature reviews and existing solutions used in the previous mobile communications, we can propose a model which manages the traffic issues starting from the point of accessing mechanism.

A Brief Description of the Proposed Model

As shown in Figure 3, the proposed model can be designed with SoDeMa for managing data traffic according to the rate, type, etc.

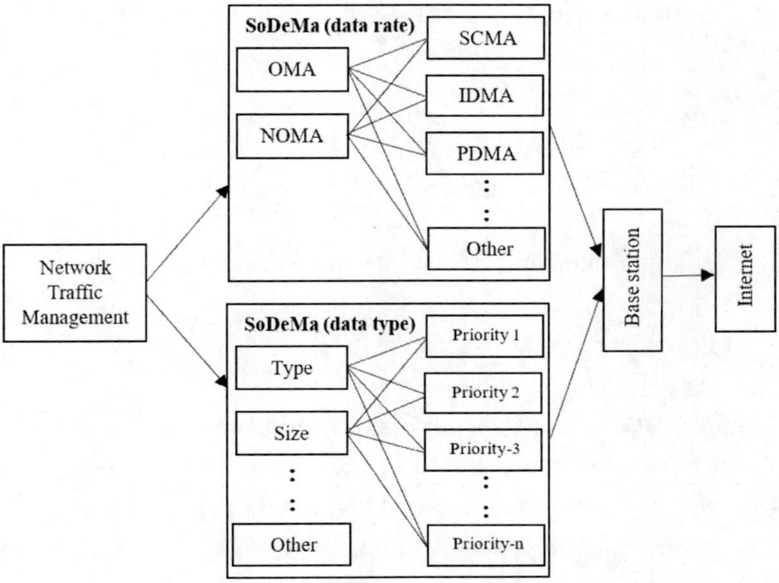

Figure 3. Traffic data management with modified slice algorithm.

Modified Slices Algorithm for Data Traffic

Although many researchers suggested the slice algorithm [23] for managing data traffic, modified slice algorithm with SoDeMa is considered for improving the 5G or 5G+ data traffic in the future. Network slicing used in the 5G applications allows the smart traffic management systems to split the data traffic according to the types of data traffic. Also, the network slice can be used to differentiate the services used in mobile applications. According to advanced communication, data traffic can be categorized into many slices such as different priorities. Each slice considered with the specific traffic protocol sent through the SoDeMa for managing data traffic efficiently with the programmable approach. Further, 5G network slicing allows users to slice the data traffic in a flexible way where SoDeMa provides efficient use of available resources within the SDN environments where network functions virtualization plays an important role.

The data traffic priorities are increasing with the situations and other factors such as sensitive. Following situation can be considered as examples of priorities.

- The medical application needs to handle data with high priority (1 ms)
- Regular monitoring activities use heavy data with medium priority (2 ms)
- Multimedia applications handle data with low priority (3 ms)

Although data traffic depends on the priorities, setting actual time needed to take action is very difficult because managing data traffic is not organized with the delay, decision making, etc. In this model, the delay is considered to arrange the priorities through the SoDeMa. End to end delay determined in the data traffic of the SDN based 5G network can be calculated using total time from application to the data layer. In (1), total delay (D) depends on the individual delay of application plane (D_a), controller plane (D_c) and data (D_d) plane.

$$D = D_a + D_c + D_d \tag{1}$$

Waiting time and intensity calculations allow 5G developers to analyze the delay and investigate the optimized value of the latency for improving 5G+ systems and requirements. Although processing time (S_c) of the controller depends on many factors such as types of data traffic, equation (2) provides us to consider the basic processing time of the controller.

$$D_c = S_c(t) \tag{2}$$

Traffic heterogeneity rate allows us to handle the traffic flow among the switches located in the data plane. Waiting time (S_{ij}) and transmission time (S_d) allow us to calculate the delays within the data plane as in (3).

$$D_d = S_d(t) + \sum_j^n S_{ij}(t) \quad (3)$$

Interpretation of studying and analyzing data traffic was very difficult due to the heterogeneity of the data traffic being handled in many services. Thus, data traffic needs to be optimized carefully for improving 5G+ requirements that heterogeneity of the data traffic can be controlled through the SoDeMa.

DATA TRAFFIC ANALYSIS

Although data traffic management depending on the technical measurements obtained from the 5G network applications, the programmable approach of SDN allows the service providers to enhance the measurement facilities. Potential data traffic carries unwanted data which needs to be identified and removed dynamically. Top of the SDN, deploying SoDeMa in the 5G system will improve traffic management and reduce the cost.

Soon massive traffic flows increase extensive pressure and volume of data on main 5G networks and all resources including spectrum. To avoid this problem, the services of the traffic should be transferred to the device to device (D2D) approach through the SoDeMa. Using D2D based traffic, we can minimize the volume of the data traffic from the main 5G network and resources used in the 5G environment. Further, service providers will reduce traffic offloading and increases spectrum efficiency.

Time Management

The traffic flows created from different sources varies with data rates depended on the number of packets transferred through the SDN channels within the given time. Managing traffic flows needs efficient technique which allows the service providers to increase the volume of the traffic within the minimum time. Handling different data rate according to the situation and priority should be organized using efficient time management. Further, SoDeMa can be deployed to distinguish the traffic flows transmitted with different data rates. In this research, delays of data and control planes depend on time management.

Comparison of Traffic

Monitoring the dynamic of Internet traffic classifies not only the types of data but also provides many benefits such as preventing traffic congestions, security analysis, etc. Although Internet traffic has been compared and analyzed by many researchers, we believe that any data traffic flow can be organized according to the priorities using SoDeMa. Traffic flow depends on the traffic volume density (bps/km^2) which is equal to totalthe data rate of all users per unit area.

Table 3.2. Statistical characteristics of traffic flow

Network traffic	Traffic Size	Packet per second Pak/sec (throughput)	Traffic fraction	Entropy of sampling
TELNET	0.920m	1022	0.77	0.337
SMTP	0.131m	145	0.11	0.362
HTTP	23m	25077	19.67	0.331
DNS	0.382m	424	0.33	0.415
ICMP	20m	22002	17.26	0.384

This network traffics given in Table 3.2 are examples of data traffic which contains statistical characteristics of traffic flows obtained from Internet applications [24]. Using SDN and SoDeMa, statistical characteristics of network traffic can be improved.

CONCLUSION

This chapter provides the details of the data traffic techniques which utilize the SDN and SoDeMa. In this efficient approach, users are able to manage the data traffic according to the service they need. Further, requirements of 5G provide the technical limitations which enhance the priority facilities. Limiting network slices and increasing EE will improve the latency which encourages users/service providers to manage the data traffic efficiently. Managing data traffic is increasing with new and/or merging technologies, devices, and applications. Therefore, we considered the basic idea of the proposed model which provides the multiple accessing facilities through the SoDeMa. Although utilizing the full or available spectrum and categorizing data traffic use the intelligent approaches, the proposed model can provide necessary priorities according to the situations intelligently with the delay needed to take action. In the future work, verifying delay calculation through the SDN planes and SoDeMa will be analyzed to improve the priorities of the data traffic in the future 5G and 5G+ environments.

REFERENCES

[1] Gamage, Samoda, Jamil Y. Khan, and Duy T. Ngo. “Adaptive Resource Allocation with Traffic Peak Duration Prediction and Admission Control for Cognitive Wi-Fi Networks.” *Computer Networks* (2018).

[2] Shen, Dawei, Wei Yan, Yuhuai Peng, Yanhua Fu, and Qingxu Deng. “Congestion Control and Traffic Scheduling for Collaborative Crowdsourcing in SDN Enabled Mobile Wireless Networks.” *Wireless Communications and Mobile Computing* 2018 (2018).

[3] Chen, Min, Yongfeng Qian, Yixue Hao, Yong Li, and Jeungeun Song. “Data-driven computing and caching in 5G networks: Architecture and delay analysis.” *IEEE Wireless Communications* 25, no. 1 (2018): 70-75.

[4] Li, Shancang, Li Da Xu, and Shanshan Zhao. "5G Internet of Things: A survey." *Journal of Industrial Information Integration* (2018).

[5] Al-Turjman, Fadi, Enver Ever, and Hadi Zahmatkesh. "Small Cells in the Forthcoming 5G/IoT: Traffic Modelling and Deployment Overview." *IEEE Communications Surveys & Tutorials* (2018).

[6] Dalla Cia, Massimo, Federico Mason, Davide Peron, Federico Chiariotti, Michele Polese, Toktam Mahmoodi, Michele Zorzi, and Andrea Zanella. "Using smart city data in 5g self-organizing networks." *IEEE Internet of Things Journal* 5, no. 2 (2018): 645-654.

[7] Dao, Nhu-Ngoc, Minho Park, Joongheon Kim, Jeongyeup Paek, and Sungrae Cho. "Resource-aware relay selection for inter-cell interference avoidance in the 5G heterogeneous network for the Internet of Things systems." *Future Generation Computer Systems* (2018).

[8] Dai, Linglong, Bichai Wang, Yifei Yuan, Shuangfeng Han, I. Chih-Lin, and Zhaocheng Wang. "Non-orthogonal multiple access for 5G: solutions, challenges, opportunities, and future research trends." *IEEE Communications Magazine* 53, no. 9 (2015): 74-81.

[9] Bogale, Tadilo Endeshaw, and Long Bao Le. "Massive MIMO and mmWave for 5G wireless HetNet: Potential benefits and challenges." *IEEE Vehicular Technology Magazine* 11, no. 1 (2016): 64-75.

[10] Agiwal, Mamta, Abhishek Roy, and Navrati Saxena. "Next-generation 5G wireless networks: A comprehensive survey." *IEEE Communications Surveys & Tutorials* 18, no. 3 (2016): 1617-1655.

[11] Guo, Yingya, Zhiliang Wang, Xia Yin, Xingang Shi, and Jianping Wu. "Traffic engineering in SDN/OSPF hybrid network." In *Network Protocols (ICNP), 2014 IEEE 22nd International Conference on*, pp. 563-568. IEEE, 2014.

[12] Zhang, Yangyang, Jianhua Ge, and Erchin Serpedin. "Performance analysis of a 5G energy-constrained downlink relaying network with non-orthogonal multiple access." *IEEE Transactions on Wireless Communications* 16, no. 12 (2017): 8333-8346.

[13] Buzzi, Stefano, Carmen D'Andrea, Tommaso Foggi, Alessandro Ugolini, and Giulio Colavolpe. "Single-carrier modulation versus

OFDM for millimeter-wave wireless MIMO." *IEEE Transactions on Communications* 66, no. 3 (2018): 1335-1348.

[14] Patcharamaneepakorn, Piya, Cheng-Xiang Wang, Yu Fu, El-Hadi M. Aggoune, Mohammed M. Alwakeel, Xiaofeng Tao, and Xiaohu Ge. "Quadrature Space-Frequency Index Modulation for Energy-Efficient 5G Wireless Communication Systems." *IEEE Transactions on Communications* 66, no. 7 (2018): 3050-3064.

[15] Liu, Jiajia, Yongpeng Shi, Lei Zhao, Yurui Cao, Wen Sun, and Nei Kato. "Joint Placement of Controllers and Gateways in SDN-Enabled 5G-Satellite Integrated Network." *IEEE Journal on Selected Areas in Communications* 36, no. 2 (2018): 221-232.

[16] Chen, Jiajia, Zhaojun Qian, Tan Wang, and Xi Li. "Modeling and simulation of IMT-2020 (5G) systems and satellite communication systems based on OPNET network simulation technology." In *Computer and Communications (ICCC), 2017 3rd IEEE International Conference on*, pp. 920-924. IEEE, 2017.

[17] Hamidi-Sepehr, Fatemeh, Ajit Nimbalker, and Gregory Ermolaev. "Analysis of 5G LDPC Codes Rate-Matching Design." In *2018 IEEE 87th Vehicular Technology Conference (VTC Spring)*, pp. 1-5. IEEE, 2018.

[18] Shu, Zhaogang, Jiafu Wan, Jiaxiang Lin, Shiyong Wang, Di Li, Seungmin Rho, and Changcai Yang. "Traffic engineering in software-defined networking: Measurement and management." *IEEE Access* 4 (2016): 3246-3256.

[19] Ding, Zhiguo, Yuanwei Liu, Jinho Choi, Qi Sun, Maged Elkashlan, I. Chih-Lin, and H. Vincent Poor. "Application of Non-Orthogonal Multiple Access in LTE and 5G Networks." *IEEE Communications Magazine* 55, no. 2 (2017): 185-191.

[20] Zhang, Yi, Huiming Wang, Tongxing Zheng, and Qian Yang. "Energy-Efficient Transmission Design in Non-orthogonal Multiple Access." *IEEE Trans. Vehicular Technology* 66, no. 3 (2017): 2852-2857.

[21] Costa-Requena, Jose, Raimo Kantola, Jesús Llorente, Vicent Ferrer, Jukka Manner, Aaron Yi Ding, Yanhe Liu, and Sasu Tarkoma.

"Software-defined 5G mobile backhaul." In *5G for Ubiquitous Connectivity (5GU), 2014 1st International Conference on*, pp. 258-263. IEEE, 2014.

[22] Pinto, Paulo, Rui Cardoso, Pedro Amaral, and Luis Bernardo. "Lightweight admission control and traffic management with SDN." In *Communications (ICC), 2016 IEEE International Conference on*, pp. 1-7. IEEE, 2016.

[23] Dighriri, Mohammed, Gyu Myoung Lee, and Thar Baker. "Measurement and Classification of Smart Systems Data Traffic Over 5G Mobile Networks." In *Technology for Smart Futures*, pp. 195-217. Springer, Cham, 2018.

[24] Wang, Jin-Fa, Hai Zhao, Shuai-Zong Si, Hao Yu, Shuai Chao, and Xuan He. *"Using Complex Network Theory for Temporal Locality in Network Traffic Flows."* arXiv preprint arXiv:1807.03454 (2018).

In: Advanced Security Issues …
Editor: Vijey Thayananthan
ISBN: 978-1-53615-538-9

Chapter 4

Analysis of Cybersecurity for 5G - Based Vehicular Communication Network

Javad Yazdani[1] and Vijey Thayananthan[2]
[1]Faculty of Science and Technology, School of Engineering,
University of Central Lancashire, Preston, UK
[2]Department of Computer Science, King Abdulaziz University,
Jeddah, KSA

Abstract

The fifth generation (5G) is going to dominate in the vehicular communication network (VCN) which handles the new communication between the driverless vehicles and other existing vehicles. In the next 50 years or more, we will have to face dual transportation systems which increase the many security risks and accidents related to unsecured communication systems employed to deliver the transportation services. In the VCN, these problems involved with cybersecurity (CS) issues between the dual transportation systems should be reduced through the appropriate security solutions. The CS issues for 5G based infrastructure incorporated with cloud and/or mobile systems will be facing many challenges such as secure transportation services. Despite many security solutions, developing group-based authentication influenced by energy-efficient cryptographic algorithms are recommended to secure the future transport systems and broadcast to achieve efficient CS objectives. Employing secure multi-level 5G based infrastructure used within the transportation systems will be an efficient method because these approaches will allow us to enhance the security features according to the evolving technologies. In VCN, appropriately applied cryptographic algorithms will be employed to improve transportation services. In this work, we prove the dynamic CS solutions which are key requirements to minimize the risks between the dual transportation systems.

Keywords: cybersecurity, 5G, VCN, transportation system

INTRODUCTION

Improving secure transportation environments in popular cities is one of the 2030 visions. In these environments, transportation services will be improved with the efficient use of technologies providing to enhance the facilities and techniques which reduce the accident rate. Here, CS issues should be identified because attacks and threats related to these issues affect the drivers' attitudes, vehicular communications, and environments. When drivers' attitudes are interrupted by these threats, the driver's behavior is affected, and it creates the uncontrollable situation. Infotainments created through the vehicular communication for good purposes and planning to continue the driving toward the destination, but wrong information is also sent to the driver by intruders, unauthorized people, and system. Environmental conditions should be checked perfectly before drivers start driving according to the situation.

Future wireless systems such as the 5th, generation (5G) of wireless and mobile networks support to design transportation services. Here, the infrastructure of 5G is being developed to reduce the cyber threats which could cause accidents. Despite the basic level of the infrastructure in the transportation systems, 5G introduces many novel approaches to improve traffic monitoring facilities through intelligent transport systems [1-2]. As shown in Table 4.1, vehicular communication functions (VCF) need to be handled securely through the appropriate security. Here, the IoT based 5G system provides us to design secure transport services via a software-defined network (SDN). As future mobile communication influenced transportation systems, a radio access network, IoT, etc. will be considered to design the secure infrastructure [3-5]. Further, technological advances such as machine to machine communication, non-orthogonal modulation, ultra-dense cells, SDN, etc. are also applicable to control the traffic services. In this research, the following strategic approaches are important and motivate us to design the efficient CS framework for securing transportation services. They are unnecessary accident rates which increase transportation management services, efficient security, and safety systems that enhance the livability of modern cities, etc. Despite these facts, threats and motivations are increasing within the transportation services because most of the vehicles are wirelessly linked. This means that cyber attacks such as Internet-borne malware can create the greatest accident risks and financial loss within the transportation organizations. In order to secure transportation services, some motivations influenced by the threats should be considered. They are malicious and nonspecific such as (malware, hacking), deliberate and planned (terrorism, criminals, etc.). As the strategic plan of the 2030 vision, transportation services should be motivated by the efficient CS issues of IoT based 5G infrastructure without affecting the legacy of transportation architecture [6-7]. Although these approaches motivate us to explore our selected research, we provide an efficient CS framework which enhances the overall security solutions with low-cost computation algorithms. Following actions may be possible to improve the transportation policies around the transportation environments. Another important point is that understanding the environmental disasters such as global warming and pollutions created from the heavy use of vehicles.

Although many possible contributions are available in the dual transportation systems, we have considered a number of points briefly in this research from the following contributions.

- Analyzing and investigating the existing CS issues which are compatible to integrate with existing transportation systems in all environmental conditions around the new and busy cities
- Investigating the collected accident details (data) using conventional and new techniques such as IoT based 5G systems and sensors. Thus, data should be collected from some selected cities where current environmental conditions and the situation of the accident should be noted.
- Designing possible CS protocols using group-based authentication influenced by energy-efficient cryptographic algorithms and developing security solutions using suitable simulation tools for improving the overall transportation systems.
- Developing a theoretical model which includes suitable CS protocols, secure 5G infrastructure, and multi-level IoT framework.
- Implementing evaluation mechanism which measures the accident rate influenced by cyber attacks within the transportation systems.

Table 4.1. Security issues

5G for transportation	Related security
IoT environments for core and VCF	VCF security (for central and distributed IoT environments)
Software Defined Mobile Network Control (SDMC)	SDN security for VCF, transport services, SDMC, etc.
Mobile network based on multi-level IoT	IoT security, network slicing security
Transport service awareness	Flexible security approach, e.g., choice of cryptographic algorithms
Adaptive allocation of functions, joint optimization of VCF and core	Applied cryptography approach, e.g., support for flexible allocation of security functions

LITERATURE REVIEW AND RELATED WORK

According to [8], developing secure network grids are paramount and priority for smart cities of the future in Indiain to improve and enhance current grids. Although the overall performance of Modern Network Grids in Smart Cities of India has acceptable levels of security and privacy, the transportation system needs extra protection with the critical success factors. In order to improve the reliability of the smart cities, a total of 16 critical success factors has been detected with the help of Principal Component Analysis.

Information technology is rapidly increasing with security solutions, but cybercrimes and threats are infecting through the potential network connection between the devices used in smart cities. Despite the digitized world, maintaining secure connections it is not always possible as the number of hackers/intruders are increasing. Authors claimed that technological advancement and greater connectivity had moved organizations to better economic markets for sustainability and to suitable security solutions. Together with better technology and greater connectivity, proper security monitoring systems and solutions for cybercrime and other CS issues can be established [9].

A Cost-Aware Hierarchical Cyber Incident Analytics (CA-HCIA) is introduced in [10] for analyzing cyber attacks. In this analysis, Monte Carlo Cyber Feature Extraction and Optimal Cost Balance Algorithms play an important role in improving the CS solutions. Further, CA-HCIA can also be applied to prevent cyber-related risks in transportation services.

According to [11], ciphertext-policy attribute-based encryption (CP-ABE) provided the necessary security solution with high computational cost. To improve the solutions, researchers minimized the computational cost and complexity issues using a new RSA-based CP-ABE scheme.

As discussed in [12], the IoT emerges with many devices connected to the Internet which provides quick access to maintain the transport services. The everyday transportation activities are to encompass everything from IoT-based services which include security and privacy. Although deployment of IoT and sensors is increasing to improve the transportation services, Figure 1 shows the VCN which provides the necessary security to enhance the transportation facilities. For instance, when service providers use secure VCN, road service units (RSU) can enhance road maintenance services more accurately.

In [13], ultra-lightweight mutual authentication approach has been proposed. Here, two bitwise operations adopted for authentication purpose, which ensures low computational and storage cost. The proposed approach resists against a different type of attacks such as; DoS, tracking, replay, etc.

Article [14] provides trust-based routing for IoT sensor based on node trustworthiness and evaluation. Here, the authors have proposed SecTrust which consists of five stages with a predefined threshold.

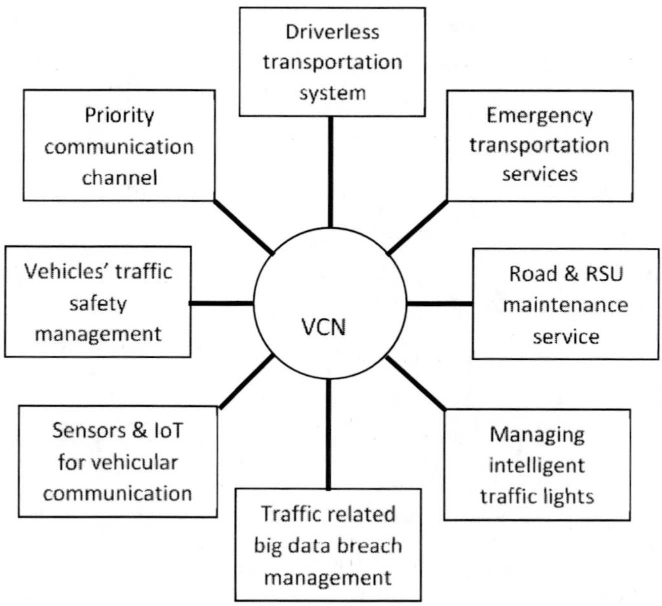

Figure 1. Secure VCN for improving transportation services.

According to [15], deploying wireless sensor networks (WSNs) allows service providers to implement the smart facilities and services between the users. Despite many sensors, WSN collects necessary data from the smart city and handles the implemented services securely. In order to improve the security around the smart city, the authors proposed a hierarchical framework which not only reduces the complexity but also improves the security. In this proposed approach, usage control (UCON) technologies play important roles to improve the security of WSNs. Using advanced persistent threat detection, features such as the continuous decision and dynamic attributes in UCON help to monitor the attacks. Further, the dynamic adaptive chance discovery mechanism was employed to detect unknown attacks.

According to [16] authors discussed how billions of Internet-connected devices and machines would change the future. Here, different communication protocols, security, privacy, etc. -in managing IoT are considered as key challenges to improving the security of massive connectivity without breaching privacy. Although IoT brings significant advantages over traditional communication technologies for smart transport and smart city applications, IoT based smart transport services and architecture yet to be implemented properly [17].

This paper [18], provides the CS of the industrial control systems (ICS) which protect not only the control systems but also all technical and management services. In their approach, the CS landscape includes 1) the key principles and unique aspects of ICS operation; 2) a brief history of cyber attacks on ICS; 3) an overview of ICS security assessment; 4) a survey of “uniquely-ICS” testbeds that capture the interactions between the various layers of an ICS; and 5) current trends in ICS attacks and defenses.

According to [19], system dynamics (SD) is a methodology that could be developed to analyze how any systems including transport systems change over time. The authors used SD based CS simulation which can imitate the attacker's activities in open system interconnected layer 4 and above. CS impact on the power grid could be useful to prevent the power controls and supply units integrated within the transportation systems [20].

Although vehicle accidents depend on many factors, analysis and decision making models helped us to reduce the accident rate through specific factor such as data quality [21]. Despite all these decision-making models, the technique of the decision modeling which increases the security and safety could be integrated with the RFID and communication networks [22, 23].

According to [24, 25], green data storage expected to use in future transportation environments, is considered with the novel cyber security solutions based on Li-Fi and quantum cryptography. Through this approach, big data used in the smart transportation systems could be secured with energy-efficient security protocols in order to improve the data traffic performance [26, 27] which is one of the evaluation mechanism used for securing data communication within the smart transport of the new smart cities.

Communications protocol design for 5G vehicular networking architecture [28, 29], allows us to improve the communication services of the transportation systems linked with the internal and external signals. To standardize the future transportation systems with new facilities such as CS issues, integrated infrastructure based on the interworking of heterogeneous technologies may be implemented within modern cities.

PROPOSED FRAMEWORK

Collecting necessary data of driverless and existing transportation systems is the problem because CS issues are growing without any proper definitions. Next few decades, we all will have to face and challenge many security problems related to the dual transportation systems. Growing technologies against evolving security issues and future CS solutions are challenging problems. Securing potential transportation systems in all popular cities in the world and monitoring data traffic related to transportation services which monitor the accident rates is also another objective of this work.

Description of the Framework

In this chapter, we develop the framework for securing 5G-based VCN through the CS solution. In the description of 5G-based infrastructure, multi-level IoT based 5G framework allows us to improve the CS solutions in

VCN. Further, the framework will enhance the protection against the evolving CS which affects future transport system. Also, it minimizes the rate of transportation accidents within the modern cities where both driverless and existing handle the secure VCN. Although new technologies minimize the potential risks and cyber attacks, evolving cyber threats should be considered within the future dual transportation system. As shown in Figure 2, security issues of multi-level IoT-based 5G infrastructure could be considered to improve the smart transportation system.

In this theoretical research, we have considered potential security challenges through the appropriate design. Development and implementation also briefly introduced for future research problem. Designing efficient CS protocols based on applied cryptography algorithms with minimum cost and maximum security solutions to improve the overall transportation systems. Developing efficient transportation model with suitable CS protocols which include the secure 5G infrastructure and multi-level IoT framework. Implementing secure transportation systems using the above objectives and evaluation mechanism which enhance the security facilities and liveability of new cities.

Methodology

As highlighted earlier, this research will encompass five inter-related research topics under the umbrella of new transportation systems design considered with the efficient solution for minimizing the accidental rate which influences the CS. The topics are: 1) Collecting necessary data of existing transportation problems and CS issues, 2) Analyzing 5G infrastructure and multi-level IoT framework of future transport system, 3) Designing efficient CS protocols based on applied cryptography algorithms, 4) Developing transportation model with CS protocols, secure 5G infrastructure and multi-level IoT framework, 5) Implementing secure transportation systems and evaluation mechanism.

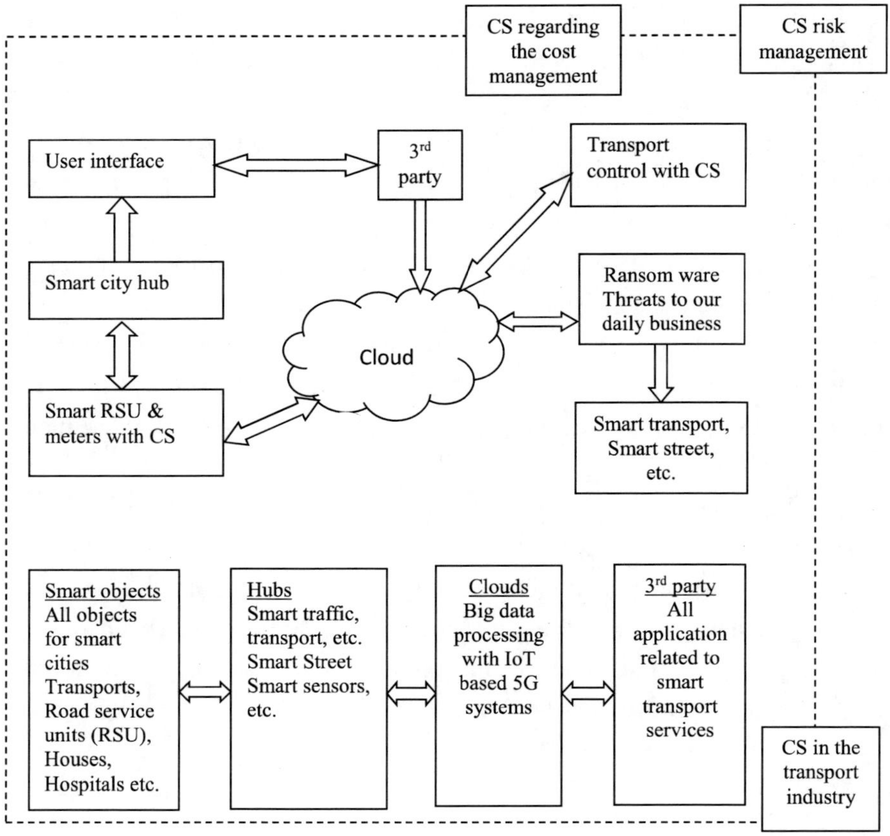

Figure 2. The framework of multi-level IoT based 5G infrastructure for the VCN.

The theoretical model is divided into a series of functions which address the key areas of CS readiness. Formulating the functions such as identity, protect, etc. for the CS solutions are part of the algorithm employed within hubs located somewhere or hidden places within the modern cities.

- *Identify* – Finding, assessing and selecting all fixed and mobile devices which are more vulnerable to attack create the CS problems, because IoT, devices, and people within the transportation environments do not know the exact identities.

- *Protect* – Implementing security controls according to the vulnerability of all components used within the modern city are more difficult and challengeable.
- *Detect* – Actively monitoring the vehicles which include the people and mobile devices is part of the functions used in the CS solutions.
- *Respond* – When an attack or incident is identified then an efficient CS solution must be ensured in order to reduce damage to the transportation environments.
- *Recover* – Most of the incidents related to cyber attacks are able to restore when above-mentioned functions are active within the proposed model.

As shown in Figure 2, security issues during the transportation services between vehicles and RSU could be simplified using multi-level IoT-based 5G infrastructure.

ANALYSIS AND DISCUSSION

Analyzing all existing theoretical model used in the security analysis allows us to improve the future security of transport systems. Further, it allows us to develop efficient CS solutions to minimize accident rates within transportation environments.

Analysis

In our theoretical analysis, controlling security levels between the driverless and existing vehicles in the next few decades will be the potential challenge. This situation will create many security questions such as passengers' safety regulations against the cyberattacks. Using functions given in the previous section, identity access management could be employed to analyze the CS levels within the transportation environments.

Identity access management is one of the components used in the IoT-based 5G infrastructure for securing management of electronic identities influenced by transport activities.

As shown in Table 4.2, security analysis within the IoT environment could be considered with the flexibility and automation of security mechanism. Here, the IoT based 5G system provides us to design the secure transport services via SDN and group-based authentication influenced by energy-efficient cryptographic algorithms.

Table 4.2. Security analysis of 5G in VCF

A higher level of security	The flexibility of security mechanisms	The possibility of security automation
Increased robustness against cyber attack	Group-based authentication influenced by energy-efficient cryptographic algorithms	VCF security orchestration and identity access management
Enhanced privacy	Vehicle user plane and VCF encryption	Self-adaptive, intelligent security controls
Cybersecurity assurance	Adjust security mechanisms per network slice	IoT based 5G approach

In the 5G security vision of the dual transportation system, a higher level of security can be considered with evolving 5G system requirements which improve not only the CS solutions but also privacy within the transportation environments. Here, the 5G network influenced by 5G systems provide the following security issues.

- Increased robustness against cyber attacks
- Enhanced privacy
- Security assurance

Vehicular communication based on 5G between the dual transportation systems should be able to handle the heavy data traffic with maximum security levels. Handling heavy data traffic needs an efficient and secure channel and power. Thus, our theoretical model allows us to modify the secure power control and communication which provide the wireless charging facilities from the future power sources such as nuclear power stations. With the 5G+ technology and CS solutions, driverless vehicles will be expecting wireless chargers which will be one of the future challenges. Although we have many applied cryptographic algorithms for securing power communication at the physical layer, CS will enhance future power communication.

Discussion

All cyber attacks related to potential accident and transportation environment could be depending on the faulty devices, human errors and section purposed human behavior.

Faulty Devices

The 5G systems have many devices communicated between the vehicles and road service units which monitor the road users and their driving behaviors. Predicting finding and detecting of faulty devices are not simple because time, situations and locations of the cyberattacks cannot be defined.

Human Errors

The compatibility of the technology issues during transportation, interaction monitoring, and unfamiliar configurations may create unexpected human errors. Cyber threats influenced by human errors also spoil the vehicular communication devices which change the security levels in many different ways.

Purposed Human Behaviors

Transportation services depend on the number of users which include the devices controlled by environmental factors such as weather conditions. Here, peoples' unacceptable behaviors and purposed attacks are serious cases of cyber threats and cyber attacks which affect transportation services and environments.

Conclusion

We have studied the security issues between the dual transportation system which deals with future driverless and all existing vehicles. According to future users and 5G-based technology, all current transportation services will be renovated offering many security monitoring facilities using group-based authentication influenced by energy-efficient cryptographic algorithms. In this research, CS solutions provide the necessary security in all transportation services. Through the theoretical studies and analysis, basic CS model is identified using multi-level IoT-based 5G infrastructure for the smart transportation system.

References

[1] Menouar, Hamid, Ismail Guvenc, Kemal Akkaya, A. Selcuk Uluagac, Abdullah Kadri, and Adem Tuncer. "UAV-enabled intelligent transportation systems for the smart city: Applications and challenges." *IEEE Communications Magazine* 55, no. 3 (2017): 22-28.

[2] Camacho, Fernando, César Cárdenas, and David Muñoz. "Emerging technologies and research challenges for intelligent transportation systems: 5G, HetNets, and SDN." *International Journal on Interactive Design and Manufacturing (IJIDeM)* 12, no. 1 (2018): 327-335.

[3] Kim, Jin Ho. "A survey of IoT security: Risks, requirements, trends, and key technologies." *Journal of Industrial Integration and Management* 2, no. 02 (2017): 1750008.

[4] Sfar, Arbia Riahi, Enrico Natalizio, Yacine Challal, and Zied Chtourou. "A roadmap for security challenges in the Internet of Things." *Digital Communications and Networks* 4, no. 2 (2018): 118-137.

[5] Samaila, Musa G., Miguel Neto, Diogo AB Fernandes, Mário M. Freire, and Pedro RM Inácio. "Security challenges of the Internet of

Things." *Beyond the Internet of Things*, pp. 53-82. Springer, Cham, 2017.

[6] Gavrilovska, Liljana, Valentin Rakovic, and Vladimir Atanasovski. "Visions towards 5G: Technical requirements and potential enablers." *Wireless Personal Communications* 87, no. 3 (2016): 731-757.

[7] Amaral, Leonardo Albernaz, Everton de Matos, Ramão Tiago Tiburski, Fabiano Hessel, Willian Tessaro Lunardi, and Sabrina Marczak. "Middleware Technology for IoT Systems: Challenges and Perspectives toward 5G." In *Internet of Things (IoT) in 5G Mobile Technologies*, pp. 333-367. Springer, Cham, 2016.

[8] Chatterjee, Sheshadri, Arpan Kumar Kar, and M. P. Gupta. "Critical success factors in establishing a 5G network in smart cities: Inputs for security and privacy." *Journal of Global Information Management (JGIM)* 25, no. 2 (2017): 15-37.

[9] Patrick, Harold, and Ziska Fields. "A need for cyber security creativity." In *Collective Creativity for Responsible and Sustainable Business Practice*, pp. 42-61. IGI Global, 2017.

[10] Gai, Keke, Meikang Qiu, and Houcine Hassan. "Secure cyber incident analytics framework using Monte Carlo simulations for financial cybersecurity insurance in cloud computing." *Concurrency and Computation: Practice and Experience* 29, no. 7 (2017): e3856.

[11] Odelu, Vanga, Ashok Kumar Das, Muhammad Khurram Khan, Kim-Kwang Raymond Choo, and Minho Jo. "Expressive CP-ABE scheme for mobile devices in IoT satisfying constant-size keys and ciphertexts." *IEEE Access* 5 (2017): 3273-3283.

[12] Mozzaquatro, Bruno A., Ricardo Jardim-Goncalves, and Carlos Agostinho. "Towards a reference ontology for security in the internet of things." In *Measurements & Networking (M&N), 2015 IEEE International Workshop on*, pp. 1-6. IEEE, 2015.

[13] Tewari, Aakanksha, and B. B. Gupta. "Cryptanalysis of a novel ultra-lightweight mutual authentication protocol for IoT devices using RFID tags." *The Journal of Supercomputing* 73, no. 3 (2017): 1085-1102.

[14] Flatt, Holger, Sebastian Schriegel, Jürgen Jasperneite, Henning Trsek, and Heiko Adamczyk. "Analysis of the Cyber-Security of industry 4.0

technologies based on RAMI 4.0 and identification of requirements." In *Emerging Technologies and Factory Automation (ETFA), 2016 IEEE 21st International Conference on*, pp. 1-4. IEEE, 2016.

[15] Wu, Jun, Kaoru Ota, Mianxiong Dong, and Chunxiao Li. "A Hierarchical Security Framework for Defending Against Sophisticated Attacks on Wireless Sensor Networks in Smart Cities." *IEEE Access* 4, no. 4 (2016): 416-424.

[16] Mohapatra, Sukant K., Jay N. Bhuyan, Pankaj Asundi, and Anand Singh. "A Solution Framework for Managing the Internet of Things (IoT)." *International Journal of Computer Networks & Communications (IJCNC)* 8, no. 6 (2016): 73-87.

[17] Stojkoska, Biljana L. Risteska, and Kire V. Trivodaliev. "A review of the Internet of Things for smart home: Challenges and solutions." *Journal of Cleaner Production* 140 (2017): 1454-1464.

[18] McLaughlin, Stephen, Charalambos Konstantinou, Xueyang Wang, Lucas Davi, Ahmad-Reza Sadeghi, Michail Maniatakos, and Ramesh Karri. "The cybersecurity landscape in industrial control systems." *Proceedings of the IEEE* 104, no. 5 (2016): 1039-1057.

[19] Kannan, Uma, Rajendran Swamidurai, and David Umphress. "System Dynamics as a Tool for Modeling Application Layer Cyber Security." In *Proceedings of the International Conference on Modeling, Simulation and Visualization Methods (MSV)*, p. 88. The Steering Committee of the World Congress in Computer Science, Computer Engineering and Applied Computing (WorldComp), 2016.

[20] Soupionis, Yannis, Roberta Piccinelli, and Thierry Benoist. "Cyber security impact on power grid including nuclear plant." In *Computer Science and Information Systems (FedCSIS), 2016 Federated Conference on*, pp. 767-773. IEEE, 2016.

[21] Thayananthan, Vijey, and Riaz Ahmed Shaikh. "Contextual Risk-based Decision Modeling for Vehicular Networks." *International Journal of Computer Network and Information Security* 8, no. 9 (2016): 1.

[22] Thayananthan, Vijey, Ahmed Alzahrani, and Muhammad Shuaib Qureshi. "Efficient techniques of key management and quantum

cryptography in RFID networks." *Security and Communication Networks* 8, no. 4 (2015): 589-597.

[23] Shaikh, Riaz Ahmed, and Vijey Thayananthan. "Hop-by-hop trust evaluation algorithm for identity anonymous sensor networks." *IETE Journal of Research* 61, no. 2 (2015): 154-159.

[24] Thayananthan, Vijey, Omar Abdulkader, Kamal Jambi, and Alwi M. Bamahdi. "Analysis of CS based on Li-Fi in green data storage environments." In *Cyber Security and Cloud Computing (CSCloud), 2017 IEEE 4th International Conference on*, pp. 327-332. IEEE, 2017.

[25] Thayananthan, Vijey, and Aiiad Albeshri. "Big data security issues based on quantum cryptography and privacy with authentication for the mobile data center." *Procedia Computer Science* 50 (2015): 149-156.

[26] Alam, Furqan, Vijey Thayananthan, and Iyad Katib. "Analysis of round-robin load-balancing algorithm with adaptive and predictive approaches." In *Control (CONTROL), 2016 UKACC 11th International Conference on*, pp. 1-7. IEEE, 2016.

[27] Hadi, Anas A., Omar A. Abdulkader, Saleh Al-Ardhi, and Vijey Thayananthan. "Analytical Model of Enhancing Traffic Performance based on Weighted Nodes." In *Computer Modelling and Simulation (UKSim), 2016 UKSim-AMSS 18th International Conference on*, pp. 337-342. IEEE, 2016.

[28] Katsaros, Konstantinos, and Mehrdad Dianati. "A conceptual 5G vehicular networking architecture." In *5G Mobile Communications*, pp. 595-623. Springer, Cham, 2017.

[29] Chiti, Francesco, Romano Fantacci, Dino Giuli, Federica Paganelli, and Giovanni Rigazzi. "Communications protocol design for 5G vehicular networks." In *5G Mobile Communications*, pp. 625-649. Springer, Cham, 2017.

In: Advanced Security Issues … ISBN: 978-1-53615-538-9
Editor: Vijey Thayananthan

Chapter 5

CYBERSECURITY ISSUES IN INDUSTRY 4.0

***Omar Abdulkader*[1]*, Kamal Jambi*[1]**
***and Alwi M. Bamhdi*[2]**
[1]Department of Computer Science,
King Abdulaziz University, Jeddah, KSA
[2]Department of Computer Science, College of Computing,
Umm AlQura University, Al-Qunfodah, KSA

ABSTRACT

Currently, we are in the era of industry 4.0 where the cyber-physical systems automate and enable data exchange among physical and virtual assets in manufacturing technologies. With the advent of the industry 4.0 revolution, new cybersecurity (CS) risks and threats are being appeared on the scene. To address the CS risks and threats in industry 4.0, we need to address all related CS issues in all technologies that enable industry 4.0. Thus, we need to investigate CS on a cyber-physical system (CPS), internet of things (IoT), machine to machine communication (M2M) and cloud computing (CC). Further, CS should be considered in different level starting from physical security issues, network security issues, application security issues, etc. According to the current technologies and evolving threats, it seems that conventional security solutions are inadequate and not suitable for industry 4.0. So, we need to have a better security solution that meets all CS requirements for all emerging technologies that enable industry 4.0. In this chapter, we aim to highlight all relevant CS issues in industry 4.0.

Keywords: cybersecurity, industry 4.0, authentication, cloud computing

INTRODUCTION

Improving secure and intelligent environments in industry 4.0 is one of the future challenges. In the current industrial revolution, governments of the popular countries plan to execute the automation and intelligent systems soon. In these environments, automation services will be improved with the efficient use of security technologies based on 5th generation (5G) and beyond.

The main purpose of this chapter is to analyze the theoretical solutions of secure industry 4.0 system using efficient CPS which not only reduce the cyber attacks and unnecessary energy costs but also increase safety system. Although the main focus is efficient CS solutions, this chapter discusses the future security solutions to detect the evolving threats in 5G+ environments of industry 4.0.

According to [13], IoT and future wireless systems such as the 5G of wireless and mobile networks improve the services through the technical challenges. Although 5G supports to develop a 5G enabled IoT (5G-IoT), and its applications, security assurance, and privacy concerns are some of the potential problems considered through the techniques using 5G-IoT. Here, the infrastructure of 5G is being developed to reduce the cyber threats which create not only the unnecessary data traffic but also damage the infrastructure of the network. Despite the basic level of the infrastructure in the industry 4.0, CPS in an automation system enhances monitoring facilities continuously. Further, 5G based on CPS introduces many novel approaches to improve the data traffic and breach monitoring facilities through intelligent automation systems. To improve the authentication, service providers employ many algorithms and protocols which are adaptable with the technologies such as dependable CC, a combination of IoT based 5G blockchain scheme, etc.

Although future mobile communication influenced with automation systems, security challenges, Radio Access Network, IoT, etc. can be considered to design the secure infrastructure for securing industry 4.0. Further, technological advances such as machine to machine (M2M) communication allow service providers to enhance the CS solutions in industry 4.0 with minimum cost and energy consumption.

LITERATURE REVIEW

According to [1], analysis of the CS of industry 4.0 technologies is important to improve the quality, efficiency, etc. of the production system used in the industry 4.0. Authors introduced the reference architecture model industry 4.0 (RAMI 4.0) which allows the service providers to analyze the CS problems related to a cloud environment.

According to [2], CS for industry 4.0 in the current literature provides potential security issues which include attacks, threats, security properties, and detection mechanism. Despite many solutions, a reference framework simplifies the security procedures with an efficient algorithm.

According to [3], using integration of CS frameworks, models and approaches, users not only get the benefits from the economic impact of IoT but also learn to develop new methods for assessing the impact of cyber risk. Although IoT provides many facilities to modern industries through the information communication technologies (ICT), it enhances the operation of the modern systems with a higher level of accessibility, integrity, availability, scalability, confidentiality, and interoperability [4].

Security challenges in cyber-physical production systems (CPPS) introduces new solutions to potential problems in industry 4.0 where many threats and risks are affecting the CPPS [5]. Further, ICT is dominating the CPPS and connecting to many services employed in industry 4.0. Although IoT plays an important role to manage most of the systems, centralized fog computing security platform designed in [6] enhances security solutions in the healthcare system.

According to [7], CS of connected autonomous vehicles needs new solutions which enhance the security levels dynamically. External attacks are major problems in autonomous vehicles. Hence authors introduced the ranking based approach which prevents the external and remote attacks. Although many firewall mechanisms are available, the software-defined networking firewall for industry 4.0 manufacturing systems [8] improves the overall performance of security levels. Temporal and spatial filtering are the components of the SDN firewall which is able to separate the manufacturing system network from another flat network.

In [9], reference architecture model industry 4.0 (RAMI4.0) is introduced for securing factories of the future (FoF). Most of the industries are merging their old and traditional approaches into industry IoT (IIoT) based solutions [12] which enhance the FoF in many ways. They are protection against evolving threats, potential security solutions, etc.

According to [11 & 15], CC enhances the IT-based manufacturing facilities in the industry 4.0 environments. In addition, CPPS integrated with cloud environment is considered with industrial internet reference architecture (IIRA) and the RAMI 4.0. The challenges of just-in-sequence supply in the automotive industry are introduced. Further, the cyber-physical logistics systems not only improve the industry 4.0 but also provide the solutions based on RAMI 4.0.

The technological advantages and challenges of basic CPS and industrial CPS (ICPS) mentioned in [16-18] are providing new management styles with solutions such as additional security. Despite many CPS solutions, ICPS enhances security management through machine learning and fog computing. Further, CPS looks after that all systems in the industry 4.0 work properly, efficiently and safely. The development of attack detection encourages industry 4.0 managers to monitor the security attacks as well as stability. In this research, the capability of weakening various attacks is one of the important features in their proposed system modeling.

CYBER THREATS AND SECURITY ISSUES

According to the literature review, the CS issues are growing with many new ideas and solutions. Although industry 4.0 technologies such as IoT, CPS, etc. are important to improve the CS solutions, evolving cyber attacks and threats in the industry 4.0 need continuous security solutions with minimum cost. Despite many CS solutions, RAMI 4.0 allows the service providers to analyze the CS problems related to a cloud environment. CS for industry 4.0 in the current environment needs multiple and continuous solutions. Potential CS issues which include cyber attacks, big data breaches, threats, etc. need a proper detection mechanism with desired security properties such as authentication.

Cyber Threats and CPS

Over the past few years, cyber-adversaries have adopted the advanced persistent threat (APT) model as part of attack campaigns. In this section, threats related to CS issues and APT are considered with examples.

In the physical devices, a novel methodology of automatic identification based on stochastic geometry approach (SGA) provides the necessary security solutions to all types of physical layer attacks. Hence, devices including IoT and M2M depend on the CPS and SGA which enhances the CS solutions to minimize all cyber attacks. We can also create an appropriate feature set for collecting the characteristics of physical layer attack and CS issues predicted in the specific and different channel conditions within the modern infrastructures of the driverless vehicles. In addition, a novel security algorithm based on intrusion detection system (IDS) is incorporated for dealing with previously unseen types of physical layer attacks. Regarding the cyber threats in the CPS, data communication influenced traffic information provision services (TIPS) needs protection for delivering proper information to drivers and driverless vehicles. Here, the progress of data communication related to vehicles and drivers can be analyzed through the data center and 3rd parties with the authorization of the TIPS. To improve the security issues in vehicular communication, TIPS should employ the CPS which enhances the security in the following ways. First, applying security protocols based on CPS for protecting the spiral information which provides the security to drivers and vehicles. Secondly, intelligent information processing which helps us to analyze the security issues in the data center and 3rd parties. Finally, information should be extracted through the secure interface which may be either human to machine or machine to machine.

Security in a Cloud Environment

Access control is one of the techniques to prevent the CS attacks and improve the authentication in CC and cloud environment.

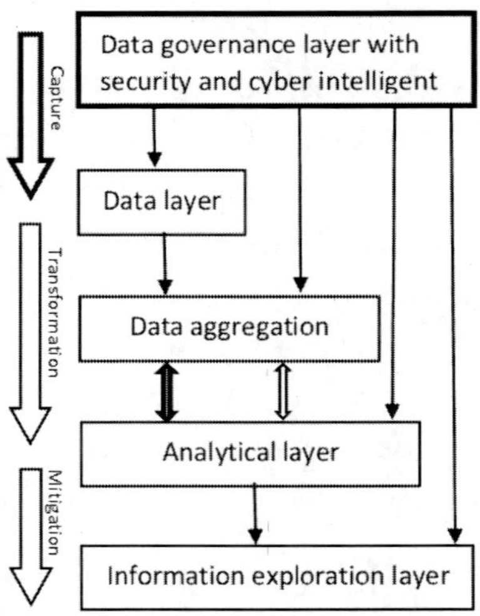

Figure 1. Analytical architecture for mitigating cyber attacks.

As shown in Figure 1, access control depends on the analytical techniques based on current security technology. This architecture illustrates the mitigation procedure of the cyber attacks through 5 layers which help us to capture the data intelligently, transform them through data and aggregation layers into analytical form and mitigate using information exploration layer. Further, this illustration can be modified to mitigate the cyber attacks from CC and cloud environment.

LAYERS OF INDUSTRY 4.0

Industry 4.0 is all about potential information and data which exchange services securely and efficiently and provide the real-time monitoring facilities between the people, machines, materials, and systems.

Functions of industry 4.0 depend on the layers shown in Figure 2 where applying security solutions with necessary cryptographic algorithms in an individual layer enhances the overall security levels. The details of security issues given in Table 5.1 allow us to improve the security solutions in industry 4.0.

CS issues should be considered in each layer and suggest the best CS solutions that meet all security requirements for all layers. Although basic desired security properties enhance the CS solutions, evolving new technologies and applied cryptography in industry 4.0 not only secure the services but also reduce the cost.

Table 5.1. Desired security properties for industry 4.0 layers

Layers	Technologies	Desired security properties
Application	CC	Confidentiality (C), Integrity (I) Authentication (A), etc.
	Big data	
Virtual	Virtual reality	Network security, CIA, etc.
Network layer	SDN	Authentication, network security Information security, etc.
	Edge computing	Wireless security, CS, etc.
	Fog computing	Cloud security, CS, etc.
Physical layer	CPS, IoT, IIoT,	CS, secrecy rate, CIA, etc.
	Sensor	Wireless security,
	Actuator	Secrecy rate, physical attack, etc.

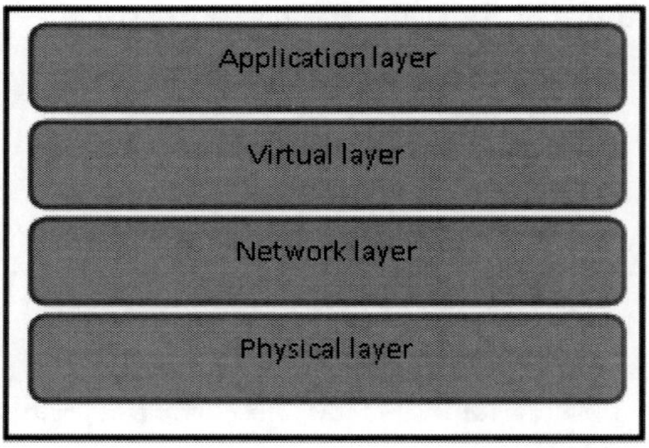

Figure 2. Basic layers of industry 4.0.

CS WITH IOT AND M2M

Security solutions based on IoT and M2M provide protected services within the industry 4.0. Although M2M is dealing with the machines, security issues between the machine type communication and protecting machine itself need new security solutions depended on the CS requirements. In cloud communication, IoT supports to enhance the secure cloud services using flexible facilities of IoT sensors and automation.

CS layers as shown in Figure 3, provide necessary security components to improve the CS solutions. CS requirements for deploying IoT and M2M in industry 4.0 considered briefly in the 5G environment, but they need to be modified for developing 5G+ and beyond. Here, we have focused on five security layers including secure protocols, security services based on IoT or M2M, CS policies, M2M gateway security, and M2M network security.

- Layer 1 deals with CS issues of internal and external attacks controlled through the appropriate protocols
- Layer 2 focuses on securing services based on M2M nodes where access control and security functions for the services are considered. CS is enhanced when system requirements of M2M are upgraded.

- Layer 3 represents the software security functions and CS policies which determine the level of the attacks when users are within the M2M network.
- Layer 4 provides the M2M gateway system security with the appropriate CS policies set by the security management influenced by service providers of M2M communication.
- Layer 5 represents the M2M network security which holds all the detection mechanism and solutions for cyber threats.

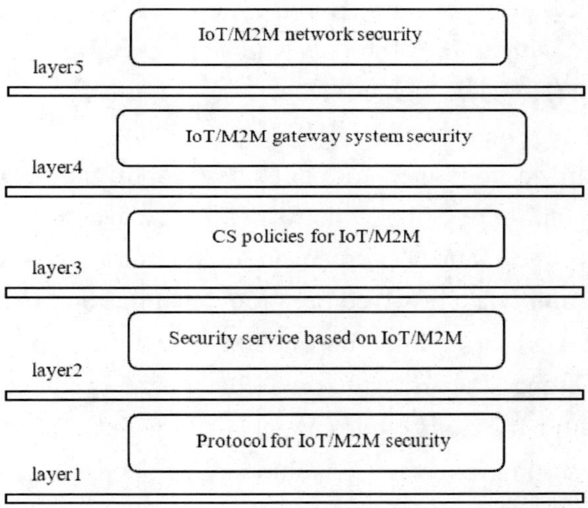

Figure 3. CS requirements for IoT and M2M communication.

Threat Model of M2M Communication Infrastructure

Wireless M2M communication is the extended version of traditional wireless communication which directly merges with the latest communication such as IoT based 5G and 5G+ communication, vehicular communication, etc.. Despite the current infrastructure with security and safety regulations, M2M communication and its infrastructure allow us to expand the CS issues which include the new threat models and dynamic solutions. Further, the legacy system not only keeps existing security systems but also it provides the interfacing facilities between the current and future infrastructure.

Authentication in Cloud

Users can protect and manage their tasks when they are in cloud environments. Regarding the authentication for dependability data, following architectures are important to which we use minimum administrations of services. In the industry 4.0, CS based on authentication can be designed with all the protocols which are TLS, Remote Access Dial-In User Service (RADIUS), Open ID, etc.

As shown in Figure 4, RADIUS, OpenID and Hybrid mode of the architecture can be combined to find the strength and weakness of the authentication. In RADIUS mode, network access and authentication servers share in solving the security problem in a cloud environment. In OpenID mode, dependability data involved between the online and OpenID services should be protected through the dependability factors. Authentications are unavoidable during the transmission that targets any two or more destinations within the cloud environments. Although CC and access control are considered to improve the CS solutions involved with authentication, following design with the different architectures will allow service providers to improve the authentication which enhances the CS solution.

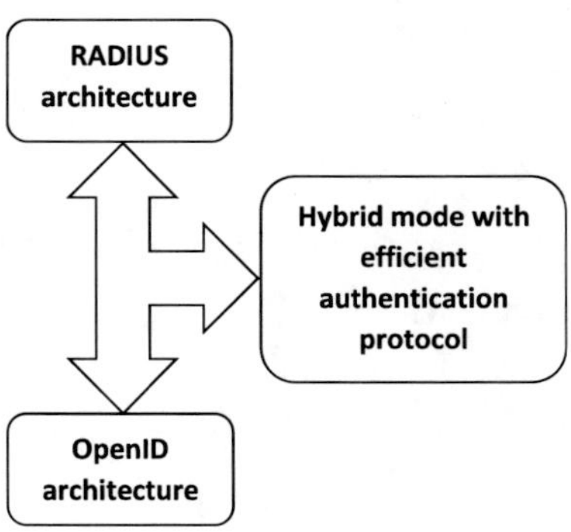

Figure 4. Authentication with different architectures.

Table 5.2. Vulnerabilities features and authentication design

Vulnerability/Supported features	RADIUS	OpenID	Hybrid
Tolerates crash faults	Yes	Yes	Yes
Tolerates arbitrary faults	No	No	No
Tolerates infrastructure outages	No	No	No
Tolerates DDoS attacks	No	No	No
Risk of common vulnerabilities	High	High	Low
Risk of sensitive data leakage	High	High	Low
Protocol security-related vulnerabilities	Yes	Yes	Yes
Susceptibility to resource depletion attacks	Yes	Yes	Yes

As shown in Table 5.2, vulnerabilities features are compared when appropriate authentication design is employed in the CS solutions which prevent not only the cyber attacks but also other industry 4.0 attacks such as DDoS.

Industry 4.0 encourages to deploy efficient authentication for improving CS infrastructure in the future industries because risk management of CS which includes data breaching, DDoS attacks, data leakage, etc. affects the product life cycles. Although many manufacturers in the industry 4.0 prefer to use the efficient design of authentication influenced to CS, they do not have enough experience and knowledge on operating environments of risk management.

Discussion of CS and Results

Regarding the CS solutions, discussing potential and future technologies are very important because many small industries face a lot of difficulties to employ CS solutions with emerging technologies. In this discussion, we have considered some key terms which affect the industry 4.0 environments. According to [14], a cyber-physical approach for integrated energy and maintenance management improve the information technology (IT) and reliability of the machines that service providers and manufacturers need.

Assuring IT security depends on the reliability and lifetime of the system created through the efficient cyber-physical approach.

As shown in Figure 5, authentication can be improved using appropriate architecture demonstrated to improve cloud security. Three different architectures are considered to measure the number of authentication per second when a number of clients are involved in cloud services. Hybrid mode of the architecture employed in Figure 4 takes a reasonable number of authentications per second. According to theory, complexity will be more in the hybrid than the other two architectures. In general, it seems that all technologies (CPS, IoT, industry 4.0, etc.) are the same in nomenclatures and implementation. Regarding the CS based on the authentication, these technologies allow us to enhance security issues. But if we analyze these technologies deeply, we can see a lot of the differences in terms of security management. CC is, in general, an IT architectural model where computing services (both hardware and software) are abstracted and delivered to customers over the Internet, on-demand, in a self-service fashion, independent of device and location.

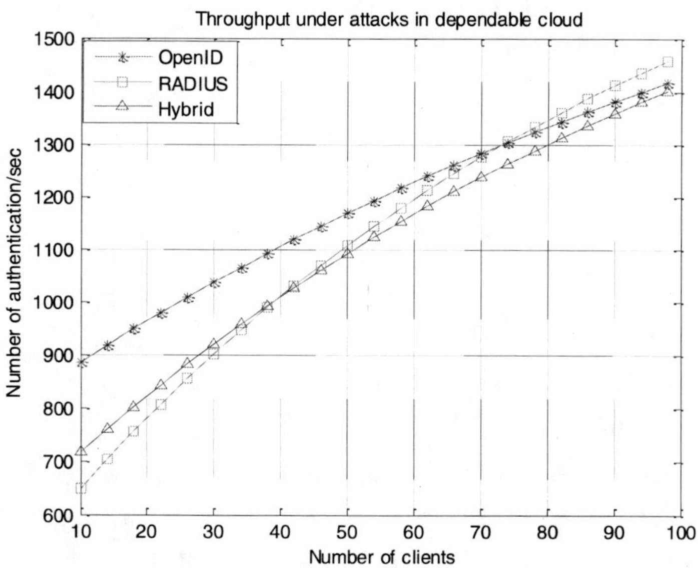

Figure 5. Comparison of authentication design for analyzing CS.

RECOMMENDATIONS

From the authentication and CS analysis of security management, the following recommendations may be possible to develop a dependable cloud system which improves CS solutions in industry 4.0.

Dynamic CS

Recent dependable cloud systems expect that dynamic CS defense can be obtained through optimized designs and key management. Even though visibility, collaboration, and workforce enrichment are essential for dynamic CS monitoring, appropriate designs allow users and service providers to improve the monitoring facilities.

New CS should be able to manage dynamically when industry 4.0 is facing real-time cyber attacks and threats. Cloud service management and design of authentication enhance the emerging facilities to provide efficient services employed in the CC infrastructure. Visibility is also controllable significantly using cyber intrusions.

Industry 4.0 encourages to deploy the dynamic CS infrastructure in the medical device industries because risk management of current CS solutions affects the product life cycles. Although many manufacturers of medical devices prefer to use the dynamic CS, they do not have enough rights which deal with the privacy issues of hospitals and their devices influenced to risk management of CS [10].

Authentication in an Enterprise Network

Enterprise network in industry 4.0 has three specific servers, network infrastructure, systems, and services. Again, authentication between the servers and users should be designed using efficient protocols that minimize the complexity, energy, and costs. To improve the CS of service users and providers in a specific organization, CS levels should be enhanced using authentication protocols. Data and service dependability provisions between the authentication gateway and server should be secured. Key management is the general solution that provides the necessary security within the cloud environment whenever security based on authentication is necessary.

Dependable Cloud Robot System

According to [19], a dependable cloud robot system can be built with the following features which are necessary to implement the secure industry 4.0. Firstly, secure automation that facilitates dynamic data collected from all resources used in the cloud environments through CPS and IoT devices. Keeping vast resources of data in onboard memory is not that easy during the transmission because maintaining onboard data not only take the unnecessary spaces but also it needs more protections such as CS solutions. Secondly, real-time multi-sensor and IoT data retrieval increase the dependable, secure services as an additional feature used in the dependable cloud robotic systems.

The cloud facilitates sharing of dependability data for securing robot learning by collecting data from IoT devices and many instances of physical trials and environments. For example, robots and automation systems based on IoT can handle many environmental conditions used in the dependable cloud system. Here, conditions may be initializing the dependable cloud system using recommended CS solutions, policies and trajectories.

Conclusion

Attacks and threats related to CS issues in industry 4.0 are studied for improving CS solutions which incorporate with the evolving technologies. In this chapter, we explored the industry 4.0 layers and security issues which include the CS solution. Also, we studied the access control scheme briefly to mitigate cyber attacks. To improve the CS solutions, we have considered the potential technologies which enhance the 5G+ schemes. Thus, 5G and

5G+ involved with IoT, M2M, CPS, and CC are considered to illustrate the CS issues in industry 4.0. In future work, we will focus on developing a practical model which will include the integration of the technologies we discussed in this chapter. Further, it will allow the service providers and users to control the CS intelligently for improving the 5G based industry 4.0 environments.

REFERENCES

[1] Flatt, Holger, Sebastian Schriegel, Jürgen Jasperneite, Henning Trsek, and Heiko Adamczyk. "Analysis of the Cyber-Security of industry 4.0 technologies based on RAMI 4.0 and identification of requirements." In *Emerging Technologies and Factory Automation (ETFA), 2016 IEEE 21st International Conference on*, pp. 1-4. IEEE, 2016.

[2] Lezzi, Marianna, Mariangela Lazoi, and Angelo Corallo. "Cybersecurity for Industry 4.0 in the current literature: A reference framework." *Computers in Industry* 103 (2018): 97-110.

[3] Radanliev, Petar, Dave De Roure, Jason RC Nurse, Razvan Nicolescu, Michael Huth, Stacy Cannady, and Rafael Mantilla Montalvo. "Integration of cybersecurity frameworks, models and approaches for building design principles for the Internet-of-Things in Industry 4.0." In: *Living in the Internet of Things: Cybersecurity of the IoT* (2018).

[4] Lu, Yang, and Li Da Xu. "Internet of Things (IoT) CS Research: A Review of Current Research Topics." *IEEE Internet of Things Journal* (2018).

[5] Kieseberg, Peter, and Edgar Weippl. "Security Challenges in Cyber-Physical Production Systems." In *International Conference on Software Quality*, pp. 3-16. Springer, Cham, 2018.

[6] Thota, Chandu, Revathi Sundarasekar, Gunasekaran Manogaran, R. Varatharajan, and M. K. Priyan. "Centralized fog computing security platform for IoT and cloud in the healthcare system." In *Exploring the Convergence of Big Data and the Internet of Things*, pp. 141-154. IGI Global, 2018.

[7] Burzio, Gianfranco, Giuseppe Faranda Cordella, Michele Colajanni, Mirco Marchetti, and Dario Stabili. "Cybersecurity of Connected Autonomous Vehicles: A ranking based approach." In *2018 International Conference of Electrical and Electronic Technologies for Automotive*, pp. 1-6. IEEE, 2018.

[8] Tsuchiya, Akihiro, Francisco Fraile, Ichiro Koshijima, Angel Ortiz, and Raul Poler. "Software-defined networking firewall for industry 4.0 manufacturing systems." *Journal of Industrial Engineering and Management* 11, no. 2 (2018): 318-333.

[9] Ehrlich, Marco, Lukasz Wisniewski, Henning Trsek, and Jürgen Jasperneite. "Modelling and automatic mapping of cybersecurity requirements for industrial applications: Survey, problem exposition, and research focus." In *2018 14th IEEE International Workshop on Factory Communication Systems (WFCS)*, pp. 1-9. IEEE, 2018.

[10] Lam, Maria Lai-Ling, and Kei Wing Wong. "Embracing cybersecurity risk management in the industry of medical devices." In *Analyzing the Impacts of Industry 4.0 in Modern Business Environments*, pp. 177-197. IGI Global, 2018.

[11] Pedone, G., and István Mezgár. "Model similarity evidence and interoperability affinity in cloud-ready Industry 4.0 technologies." *Computers in Industry* 100 (2018): 278-286.

[12] Sisinni, Emiliano, Abusayeed Saifullah, Song Han, Ulf Jennehag, and Mikael Gidlund. "Industrial Internet of Things: Challenges, Opportunities, and Directions." *IEEE Transactions on Industrial Informatics* (2018).

[13] Li, Shancang, Li Da Xu, and Shanshan Zhao. "5G internet of things: A survey." Journal of Industrial Information Integration (2018).

[14] Neef, Benjamin, Christopher Schulze, Gerrit Posselt, Christoph Herrmann, and Sebastian Thiede. "Cyber-physical Approach for Integrated Energy and Maintenance Management." *In Eco-Factories of the Future*, pp. 103-125. Springer, Cham, 2019.

[15] Juhász, János, and Tamás Bányai. "What Industry 4.0 Means for Just-In-Sequence Supply in Automotive Industry?" In-*Vehicle and Automotive Engineering*, pp. 226-240. Springer, Cham, 2018.

[16] O'Donovan, Peter. *An industrial analytics methodology and fog computing cyber-physical system for Industry 4.0 embedded machine learning applications.* PhD Thesis, University College Cork (2018).

[17] Ding, Derui, Qing-Long Han, Yang Xiang, Xiaohua Ge, and Xian-Ming Zhang. "A survey on security control and attack detection for industrial cyber-physical systems." *Neurocomputing* 275 (2018): 1674-1683.

[18] O'Donovan, Peter, Colm Gallagher, Ken Bruton, and Dominic TJ O'Sullivan. "A fog computing industrial cyber-physical system for embedded low-latency machine learning Industry 4.0 applications." *Manufacturing Letters* 15 (2018): 139-142.

[19] Wang, Lujia, Ming Liu, and Max Q-H. Meng. "Real-time multisensor data retrieval for robotic cloud systems." *IEEE Transactions on Automation Science and Engineering* 12, no. 2 (2015): 507-518.

In: Advanced Security Issues …
Editor: Vijey Thayananthan

ISBN: 978-1-53615-538-9

Chapter 6

EFFECTIVE BANDWIDTH MANAGEMENT FOR MITIGATING DOS ATTACK IN SDN

Muhammad Aminu Lawal
Department of Computer Science,
King Abdulaziz University, Jeddah, KSA

ABSTRACT

Software-Defined Networks (SDN) is a modern network architecture that offers centralized management through a controller by decoupling the control plane from the data plane. However, by employing a central controller, malicious users have a strategic point to attack the networks, an example of attack executed is denial of service (DoS). In this chapter, a bandwidthGuard (BWG) is proposed to effectively manage the bandwidth between the OpenFlow switches and the SDN controller. The scheme allocates bandwidth to users according to Quality of Services (QoS) requirements and then sets a threshold. Every violation of the threshold prompts the mechanism to half the bandwidth and set a countdown timer. This mechanism will reduce the effect of the attack and protect resources from depletion. BWG can be implemented on mininet. The scheme requires fewer resources to operate and will have a high response time which will translate to low overhead. In addition, it can easily be integrated into any controller.

Keywords: software defined network, denial of service, bandwidth management, quality of services

INTRODUCTION

The emergence of the SDN has simplified the management of the computer networks; this simplicity is provided through the separation of control plane and data plane. An intelligent central controller that has an overview of the network is employed to perform the forwarding decisions [1].

An OpenFlow protocol is utilized by the controller to effectively communicate with devices in performing the required forwarding decisions [1]. However, employing a central controller has provided malicious users a strategic point to attack the networks, an example of attack executed is DoS, which is used to overwhelm or compromise the controller with high number packets. Hence, resources such as bandwidth are consumed, and subsequently, the network is paralyzed.

Recently, several authors have proposed solutions to secure the computer and communication network using DoS [2 - 4]. Majority of these solutions concentrated on the DoS related to the packet-in messages which saturated the OpenFlow switches, thereby depleting resources. However, as a better solution, the flow level bandwidth control is considered in this research.

In this chapter, a bandwidth-Guard is proposed to effectively manage the bandwidth linking with the SDN controller and the OpenFlow switch. The mechanism allocates bandwidth to users according to QoS requirements and then sets a threshold. Every violation of the threshold prompts the mechanism to half the bandwidth and set a countdown. This approach will reduce the effect of the attack and protect resources from depletion.

The rest of the chapter is organized as follows: section 2 provides background of SDN, SDN reference model, SDN workflow and DoS attack, as well as the threat model, section 3 the related work and comparison between DoS mitigation schemes in SDN, are discussed, the proposed solution and attack analysis are discussed in section 4. Finally, section 5 concludes the paper.

Background

In recent time, SDN has emerged as a rising network management paradigm. The Open Networking Foundation (ONF) [5], a group devoted to standardization, development, and commercialization of SDN has given an obvious meaning of SDN below:

> "SDN is an emerging network architecture where network control is decoupled from forwarding and is directly programmable."

According to the definition, separating of control and data plane as well as programmability on the control plane are the main features of SDN. However, both characteristics are not new in the networking landscape. Firstly, numerous prior studies have made attempts to foster the ability to program the network. An example is efforts made to employ a software system to control the network in real-time. In [6], a network solution called SwitchWare, which permits adjusting the network operation automatically by packets flowing through the network was proposed. In the same way, software routing suites on traditional personal computer hardware like Click [7], XORP [8], Quagga [9], and BIRD [10] made efforts to develop programmable routers such that changing the operation of the network can be achieved by modifying the existing software.

Secondly, in the previous decade, decoupling control and data planes have increased. According to [11], the routing control platform (RCP) was proposed. In the RCP, the Border Gateway Protocol is interchanged with central routing control to decrease complication of the entire distributed path computation. Similarly, A Forwarding and Control Element Separation (ForCES) framework was released for improving protocol. It splits packet forwarding and control elements in ForCES Network as given in [12]. Further, a 4-D scheme was proposed in [13]. It consists of architecture with four levels namely “Decision,” “Dissemination,” “Discovery,” and “Data” arranged in the mentioned order, respectively.

More recently, the splitting of the data-control plane has been detailed. Companies such as Cisco have implemented the decoupling of the data-control plane in their devices such as Nexus 7000 series switches and Cisco ASR 1000 series routers. The planes are manufactured in modules to permit the existence of both functioning control plane and redundant plane for effective fault tolerance as well as transparently upgrading the software. The SDN perspective and its distinctiveness exist through its ability to deliver programmability by separating control and data planes. Particularly, the SDN architecture delivers the ability to program network equipment easily instead of manufacturing complex equipment as obtainable in networks before SDN. Furthermore, SDN suggests a network architecture that splits of control and data planes. By applying this architecture, data flows will not be affected by network control functions. Therefore, controllers can be intelligent as well as control the switches while the switching devices can deal with data forwarding. The separation of data plane from the control plane offers the opportunity to an external software towards configuring network behavior.

SDN Reference Model

The SDN model was advocated by ONF, as shown below in Figure 1. It comprises an application layer, a control layer, and an infrastructure layer.

The switching equipment such as switches, routers, etc. resides on the infrastructure layer of the data plane. The devices perform two functions. Firstly, they are in charge of aggregating network status, keeping the status report briefly in local devices and forwarding the report to controllers. These report can contain details like network topology, traffic statistics, as well as network usages. Secondly, the equipment implement rules are given by the controller in treating packets [1].

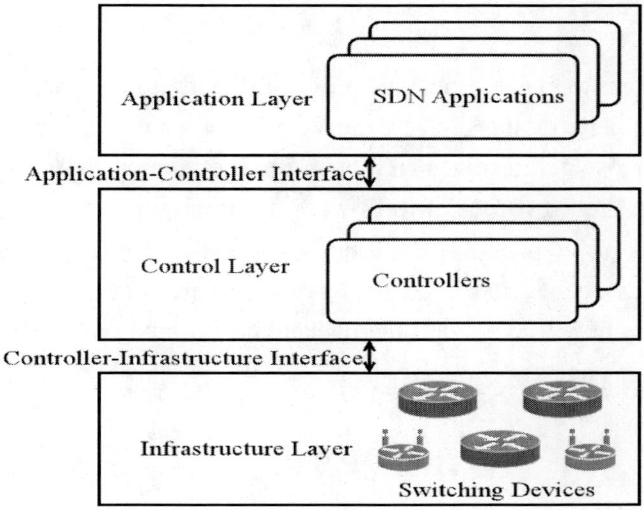

Figure 1. SDN Architecture [1].

Application layer and infrastructure layer are connected by control layer through the south-bound interface (communication in the downward direction with the infrastructure layer) and north-bound interfaces (communication in the upward direction with the application layer). The south-bound interface stipulates the functions of the switching device for the controllers which are packet forwarding rules and network status reports. The north-bound interfaces offer the application programming interface utilized by SDN application to obtain reports (network status) as provided by the switching devices; it makes system modification decisions according to the information as well as carrying out the changes by modifying the forwarding rules for the switching equipment [1].

The application layer consists of applications conceived to satisfy the requirements of the users. Using the control layer's ability to be programmed, SDN applications are capable of accessing as well as controlling devices residing on the infrastructure layer. Dynamic access control, network virtualization, server load balancing, seamless mobility, and migration are examples of SDN applications [1].

Workflow of SDN Network

In SDN networks, the responsibility of dictating the rules of operation on the entire network lies on the controller through defining data plane flow rules. When a switch receives an inbound packet, the packets are processed according to the flow rules of the switch. In the event of table miss, i.e., if the flow rule of the packet is not found in the table. The switch generates a packet-in message with a packet header of the controller which in turn makes a decision on how to deal with the packet and respond with a flow rule for the respective packet.

DoS Attack

A DoS attack in OpenFlow SDN networks means stretching computing resources in the network such that the switching devices fail to perform their required functionality, i.e., packet forwarding. An attack can be carried out by sending a huge amount of fake packets in the direction of the switching device in order to initiate a table miss which in turn will generate packet-in messages in the direction of the controller. This action will consume the bandwidth linking the controller and the switching equipment and prevent legitimate packets from being processed.

Threat Model

A DoS attack can take place in SDN networks; a user can have control over more than one computer to generate and sends a high number of packets. The attacker or malicious user doesn't need any special privileges to execute such attacks. The threat model in Figure 2 shows how an attacker can perform a DoS attack. The figure consists of a controller, an OpenFlow switch and two users (a legitimate and illegitimate). The illegitimate user or the attacker generate and send a massive number of packets with spoofed IP addresses and normal traffic to the OpenFlow switch, this action will cause table miss due to the fake packets. The table miss will prompt the switching equipment to send packet-in messages to the controller in order to acquire flow rules message for the table miss packets. Since this packets don't have flow rules, these messages will exhaust the bandwidth linking the controller and the switching equipment, memory of the switching equipment as well as computational resources will be overwhelmed by the messages. Hence a DoS attack.

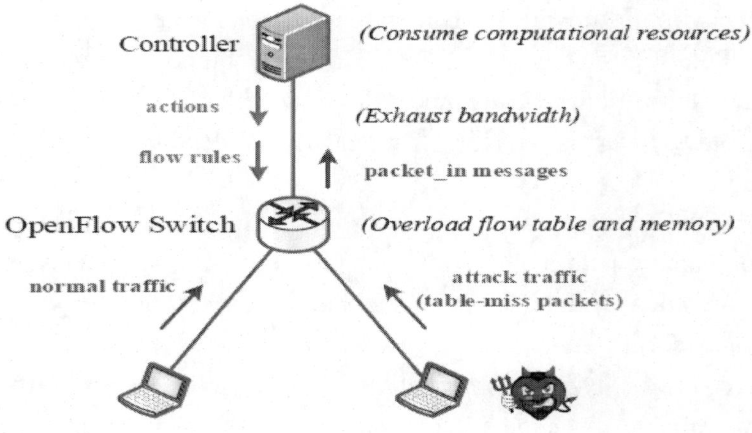

Figure 2. Threat model of DoS attack in SDN [2].

Related Work

Several mechanisms have been proposed to mitigate the DoS attacks on the SDN controller; the mechanisms are reviewed below.

In [2], a FloodDefender was proposed; it sits in the middle of the controller applications and platform. The FloodDefender comprises four modules, which are attack detection, table-miss engineering, packet filter, and flow management. Firstly, the attack detection module observes some of the network status parameters like CPU, memory, and packet-in message rate according to the research given in [6] and activate other modules if the attack is detected. Secondly, the table-miss engineering module diverts table miss packets using protecting forwarding rule to neighbor switching devices to protect the target switch from resource exhaustion (bandwidth). Then packet filter module functions as a low-level application in between the applications and the controller to recognize the illegitimate traffic and removes it to guard against control plane computational resources depletion. Finally, monitoring rules are installed and managed on the target switching device by the module in charge of flow management. The FloodDefender was implemented in software and hardware, the software modules were developed as applications on RYU controller [14] and emulated using the mininet environment while the hardware was implemented on an OpenFlow switch, Polaris xSwitch X10-24S2Q [15]. The FloodDefender enhances flow table utilization, reduces time delay and packet loss rate; in addition, the FloodDefender is scalable, can easily be deployed without extra hardware.

A FloodGuard mechanism was proposed in [3], it consists of two modules namely packet migration and proactive flow rule analyzer. The packet migration module detours the table-miss packets to the cache on the data plane during an attack, i.e., when there is packet saturation. The proactive flow analyzer automatically trails the runtime values of the state sensitive variables from the executing apps, change the obtained path states to the pro-active flow rules automatically and subsequently the rules are installed on the switching devices. Finally, the table-miss packets are sent to the controller as a packet in messages by employing the round-robin scheduling algorithm and rate limiting. The FloodGuard was implemented in software and hardware. In software, the modules are developed as applications on POX controller and emulated on mininet environment. In hardware, an OpenFlow enables commercial LinkSys WRT54GL switch with Pantou [16]. Results show that FloodGuard, i.e., effective and can be deployed with minor system overhead.

A Flowsec was proposed in [4], it utilizes a Meter proposed in [17], which has the ability to control and calculate the rate of packet transmission. The meter activates a restriction when the transmission rate of packets exceeds a predefined limit. Flowsec is implemented on the Floodlight controller.

A Port Hopping Based DoS Mitigation (PH-DM) mechanism was proposed in [18], the port hopping functionality unit of the server is dispatched to the controller; it decreases the big load due to server-side port hopping. The controller in collaboration with the OpenFlow switching device serves as a gateway to protect against DoS attack; it examines clients connecting to the network to verify if the destination ports are the same with the port opened in the established connection during the existing time slot. If the ports tally, then flow rules conforming to the switching devices transmission route are installed, and packets are transmitted to the intended server if the flow rules correspond. This ensures successful communication between legitimate clients and the server. However, if the destination port is not the same as the port opened, the connection is aborted. PH-DM is implemented on the NOX controller and simulated on mininet. The mechanism can efficiently mitigate DoS attack as well as port scanning with minimum resources of the SDN controller.

In [19], a collaborative detection and containment scheme is proposed. The scheme consists of two vital components, which are monitor and correlator, the correlator is located on the SDN controller. The monitor has two functions, which are sensing by employing an IDS and a messenger to send warning signals as well as supplementary information on the possible attack to the correlator. The received warning from the monitor is then compared with open virtual switch info by the correlator through deep inspection. It confirms if an attack is carried out on the network, identify the source of the attack traffic and mitigates it by dropping the flow. The scheme is implemented in the Global Environment for Network Innovations (GENI) cloud infrastructure. The scheme is scalable; it also has the ability to handle large traffic as well as attacks on a large scale.

In [20], SDN-Guard was proposed; it maintains communication with an IDS, which inspects packet-in messages and send alerts concerning the threat level of each flow. The scheme comprises three components, which are flow management, rule aggregation and monitoring. The flow management component determines the route of every flow as well as strict timeout consistent with Ternary Content Addressable Memory (TCAM) of the switching device according to the threat level information of the flow in order to guard against the effect of DoS attack. The rule aggregation component is responsible for gathering the entries of malicious flows so as to decrease the entries utilized in switching device TCAM. The monitoring component is in charge of gathering various information about traffic, switching devices and bandwidth, which will be utilized by the earlier mentioned components. SDN-Guard is implemented on the Floodlight controller and emulated on mininet. The scheme minimizes effects of DoS attacks by considerably decreasing controller in-bound throughput as well as consumed controller bandwidth; it also reduces the memory used by the switches. In addition, the scheme ensures packet loss rate, as well as average round trip time, decreases even under DoS attacks.

In [21], LineSwitch was proposed to mitigate the saturation attacks on the control plane. It merges network traffic blacklisting based on probability and SYN proxy techniques. LineSwitch proxies the entire inbound TCP connection originating from an IP till one is established, the rest of the following of the SYN packets are proxied with a minimal probability. It is implemented on POX controller and emulated on mininet. LineSwitch decreases time overhead while guaranteeing protection.

In [22], SGuard scheme was proposed, it consists of 2 modules namely Access control and classification modules. The module in charge of access control gathers appropriate information like ID of the switch, IP, MAC or port number; the information is joined together in a hash table. Through checking the source MAC/IP and looking across the Hash table, the scheme makes the decisions on how to treat the packet, either to forward or to drop the respective spoofed packet. Still, in an event where the recognized attackers keep sending the packets to the controller, the switching device will drop the packets. The classification module comprises of the data collector, feature extractor and classifier, which operates on the control platform as an application. Data collector intermittently demands traffic entries stored in the flow tables of switching devices using a secure channel. Then, Feature extractor instantly obtains the current characteristics of the flow entries. Finally, the classifier module uses Self Organizing Maps (SOM) in categorizing the flows as either usual traffic or malicious traffic. It also uses cache on the data plane to briefly stores table-miss packets when a saturation attack occurs. SGuard is implemented on the NOX controller and emulated on mininet. The mechanism is lightweight, efficient and incurs reasonably low overhead.

In [23], SDN Manager, a lightweight and fast DoS detection and mitigation system was proposed. It comprises of five components namely monitor, forecast engine, checker, updater, and storage service. It operates in three states: observed state, proposed state (PS) and target state (TS). In observed state (OS), monitor periodically collects an up-to-date current view of the real network states, including flow rules statistics and path conditions, from switching devices, and converts them to OS variables. The forecast engine employs a novel dynamic time-series model which greatly improves bandwidth prediction accuracy. It Reads these variables and forecasts the expected bandwidth utilization for each flow based on its historical data trace. Using a bandwidth checker module, Flows with bandwidth consumption higher than the predicted bandwidth consumption will be penalized by the application. SDN Manager merges these PS, the expected bandwidth utilization) into a TS that is assured of preserving the security as well as the performance of the system. Updater module subsequently updates the TS of the network. Storage service as the center of the system continually saves the variables of observed state, proposed state and target state and provides a highly accessible, read-write interface to other components. SDN Manager is implemented on the Floodlight controller. The mechanism is efficient while incurring a negligible overhead.

Table 1 presents a comparison table of DoS mitigation schemes in SDN based on proposed scheme, the controller used for implementation, an evaluation tool used, performance metrics evaluated, other schemes used as a benchmark for evaluation and their strength are provided. The table shows that several controllers can be employed for the implementation of the DoS mitigation schemes in SDN. However, the most popular controllers utilized are POX, NOX and Floodlight controller.

Table 1. Comparison of DoS mitigation approaches in SDN based schemes

S/No	Proposed Scheme	Controller	Evaluated Against	Performance metric Evaluated	Simulation tool	Strength
1	FloodDefender [2]	RYU	• FloodGuard • OpenFlow	• Available bandwidth • CPU utilization. • Packet Loss Rate	Mininet	• Enhances packet loss rate, time delay, and flow table utilization. • The scheme is scalable.
2	FloodGuard [3]	POX	• OpenFlow	• Bandwidth • CPU utilization. • Delay	Mininet	• Effective and can be deployed with minor system overhead.
3	FlowSec [4]	Floodlight	• With and without a proposed mechanism	• Total Traffic in Bits per sec	Not mentioned	• Mitigate DoS
4	PH-DM [18]	NOX	• With port hopping and without hopping	• Response time • Transmission success rate	Mininet	• Effective against DoS attack and • Effective against port scanning.

S/No	Proposed Scheme	Controller	Evaluated Against	Performance metric Evaluated	Simulation tool	Strength
5	A collaborative detection and containment scheme [11]	SDN controller	• Nil	• Time	GENI	• Processing high volume of traffic, • Handles huge scale attacks.
6	SDN-Guard [20]	Floodlight	• With and without SDN-Guard	• Throughput • Average round trip time • Average table size	Mininet	• Minimize the impact of DoS.
7	LineSwitch [21]	POX	• Avant-Guard	• Saturation time	Mininet	• Reduces the time overhead.
8	SGuard [22]	NOX	• With and without SGuard	• CPU utilization • Cache utilization • Delay	Mininet	• Lightweight, efficient and with relatively low overhead.
9	SDN Manager [23]	Floodlight	• SGuard • FloodGuard	• Bandwidth • CPU utilization • Response time	Mininet	• Negligible overhead.

Similarly, mininet is the most utilized evaluation tool employed to test the mitigation schemes in an SDN environment as shown on the table. Performance metric evaluated include CPU utilization, bandwidth utilization and response time which are about the resources in the network environment. For any mitigation scheme to be effective the metric showing the consumption of resources should be kept in balance. Majority of the schemes performed well based on the strengths provided. The schemes achieved minor overhead. Concerning benchmarks, most of the schemes are evaluated with the implemented scheme and without to show its effectiveness on the network. Some schemes are evaluated against each other which show continuity in the research.

PROPOSED BANDWIDTHGUARD SCHEME

The proposed BWG aims to prevent the impact of DoS attack by monitoring bandwidth consumption and applying protective measures in the event of a perceived attack. Figure 3 below shows the BWG model.

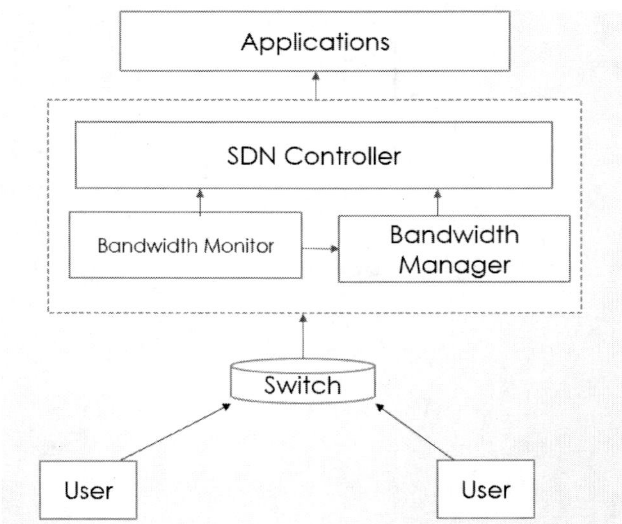

Figure 3. The BWG model.

The proposed scheme consists of two modules namely Bandwidth monitor and Bandwidth manager. The bandwidth monitor allocates bandwidth (B) to users according to their QoS requirements, set a threshold for the bandwidth and also monitors the bandwidth consumption. When the threshold is violated, the bandwidth manager reduces the bandwidth of the traffic source port by half and set a countdown timer. This process is performed anytime the threshold is violated and continues until the bandwidth consumption is within the defined limit shown in equation 2.

Given the threshold (*the*), the traffic flow (T) and bandwidth (B),

If,

$$(T \geq th) \tag{1}$$

Then,

Bandwidth is $\frac{B}{2}$ continuously until,

$$(T \leq th) \tag{2}$$

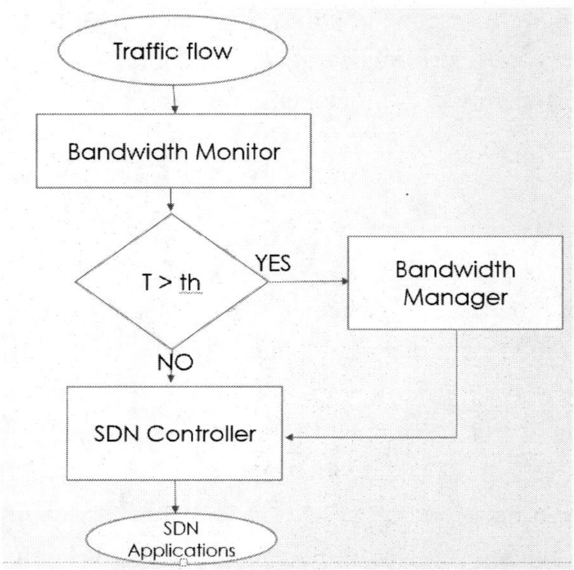

Figure 4. The flow chart of bandwidth card.

The scheme will also reduce the traffic flow to the controller and give the switch time and space to treat the table miss packets. BWG minimizes the effect of the attack as well as allow legitimate traffic to pass. Figure 4 below illustrates the workflow of the proposed scheme.

The traffic flow (T) is monitored by the bandwidth monitor and check if the consumption is within limit, if yes the traffic is forwarded to the controller, and if the consumption is above the threshold the traffic is redirected to the bandwidth manager, it halves the bandwidth for every instance of violation to neutralize the effect of overwhelming bandwidth consumption before forwarding the traffic to the controller.

Methodology

The mininet environment provides a suitable landscape to evaluate SDN. It enables researchers to deploy large networks on a single computer with limited resources. Also, it offers convenience and practicality at a low cost in contrast to test beds which are expensive and not flexible. It offers different topologies for evaluating SDN. Mininet is easy to use, scalable and accurate in term of performance [24]. The proposed scheme can be implemented efficiently in a mininet environment.

Attack Analysis

The major goal of the scheme is to resist illegitimate bandwidth consumption which overloads the bandwidth and consequently the SDN controller with heavy traffic flow to process. BWG is capable of protecting the SDN from attacks such as flow table overloading, control plane saturation attack and SYN flooding attacks as well as buffer saturation attacks. The mention attacks all utilize the available bandwidth in execution. The proposed scheme mitigates such threats effectively. In addition, the scheme can easily be integrated into any controller due to its simplicity.

CONCLUSION

In this chapter, a bandwidth managing scheme "BWG" was proposed to detect and prevent DoS attack in SDN. The scheme allocates bandwidth to users according to QoS requirements and then sets a threshold. Every violation of the threshold prompts the mechanism to half the bandwidth and set a countdown. This approach allows users to reduce the effect of the attack and protect resources from depletion. BWG can be implemented on mininet. The scheme requires fewer resources to operate and will have a high response time which will translate to low overhead and can easily be integrated into any controller. In addition, a comparison of proposed DoS mitigation schemes in SDN is presented.

REFERENCES

[1] Xia, Wenfeng, Yonggang Wen, Chuan Heng Foh, Dusit Niyato, and Haiyong Xie. "A survey on software-defined networking." *IEEE Communications Surveys & Tutorials* 17, no. 1 (2015): 27-51.

[2] Shang, Gao, Peng Zhe, Xiao Bin, Hu Aiqun, and Ren Kui. "FloodDefender: Protecting data and control plane resources under SDN-aimed DoS attacks." In *IEEE INFOCOM 2017-IEEE Conference on Computer Communications*, pp. 1-9. IEEE, 2017.

[3] Wang, Haopei, Lei Xu, and Guofei Gu. "Floodguard: A dos attack prevention extension in software-defined networks." In *2015 45th Annual IEEE/IFIP International Conference on Dependable Systems and Networks*, pp. 239-250. IEEE, 2015.

[4] Kuerban, Mutalifu, Yun Tian, Qing Yang, Yafei Jia, Brandon Huebert, and David Poss. "FlowSec: DOS attack mitigation strategy on SDN controller." In *2016 IEEE International Conference on Networking, Architecture and Storage (NAS)*, pp. 1-2. IEEE, 2016.

[5] Ali, Syed Riffat. "Software-Defined Networking (SDN)." In *Next Generation and Advanced Network Reliability Analysis*, pp. 105-130. Springer, Cham, 2019.

[6] Alexander, D. Scott, William A. Arbaugh, Michael W. Hicks, Pankaj Kakkar, Angelos D. Keromytis, Jonathan T. Moore, Carl A. Gunter, Scott M. Nettles, and Jonathan M. Smith. "The SwitchWare active Network Architecture." *IEEE Network: The Magazine of Global Internetworking* 12, no. 3 (1998): 29-36.

[7] Kohler, Eddie, Robert Morris, Benjie Chen, John Jannotti, and M. Frans Kaashoek. "The Click modular router." *ACM Transactions on Computer Systems (TOCS)* 18, no. 3 (2000): 263-297.

[8] Handley, Mark, Orion Hodson, and Eddie Kohler. "XORP: An open platform for network research." *ACM SIGCOMM Computer Communication Review* 33, no. 1 (2003): 53-57.

[9] Chen, Yu, Yifeng He, Zhengquan Zhao, Xuanhan Liang, Qimei Cui, and Xiaofeng Tao. "A Quagga-Based OSPF Routing Protocol with QoS Guarantees." In *2018 24th Asia-Pacific Conference on Communications (APCC)*, pp. 5-6. IEEE, 2018.

[10] Makowski, Łukasz, and Paola Grosso. "Evaluation of virtualization and traffic filtering methods for container networks." Future Generation Computer Systems 93 (2019): 345-357.

[11] Feamster, Nick, Hari Balakrishnan, Jennifer Rexford, Aman Shaikh, and Jacobus Van Der Merwe. "The case for separating routing from routers." In *Proceedings of the ACM SIGCOMM workshop on Future directions in network architecture*, pp. 5-12. ACM, 2004.

[12] Wang, Weiming, Ligang Dong, Bin Zhuge, Ming Gao, Fenggen Jia, Rong Jin, Jin Yu, and Xiaochun Wu. "Design and implementation of an open programmable router compliant to IETF ForCES specifications." In *Sixth International Conference on Networking (ICN'07)*, pp. 82-82. IEEE, 2007.

[13] Yan, Hong, David A. Maltz, TS Eugene Ng, Hemant Gogineni, Hui Zhang, and Zheng Cai. "Tesseract: A 4D Network Control Plane." In *NSDI*, vol. 7, pp. 27-27. 2007.

[14] Sheikh, Mohammad Nowsin Amin. "SDN-Based Approach to Evaluate the Best Controller: Internal Controller NOX and External Controllers POX, ONOS, RYU." *Global Journal of Computer Science and Technology* (2019).

[15] Polaris networks co. ltd. "Polaris XSwitch." http://www.polarisdn.com/%0Aen/product/html/?80.html.

[16] Y. Yiakoumis, J. Schulz-Zander, and J. Zhu. 2011. "Pantou : OpenFlow 1.0 for OpenWRT."

[17] *OpenFlow Switch Specification.* 2015. https://www.opennetworking.org/wp-content/uploads/2014/10/openflow-switch-v1.5.1.pdf.

[18] Zhang, Liancheng, Yi Guo, Huiqiang Yuwen, and Yu Wang. "A port hopping based DoS mitigation scheme in SDN network." In *2016 12th International Conference on Computational Intelligence and Security (CIS)*, pp. 314-317. IEEE, 2016.

[19] Chin, Tommy, Xenia Mountrouidou, Xiangyang Li, and Kaiqi Xiong. "Selective packet inspection to detect DoS flooding using software defined networking (SDN)." In *2015 IEEE 35th International Conference on Distributed Computing Systems Workshops*, pp. 95-99. IEEE, 2015.

[20] Dridi, Lobna, and Mohamed Faten Zhani. "SDN-guard: DoS attacks mitigation in SDN networks." In *2016 5th IEEE International Conference on Cloud Networking (Cloudnet)*, pp. 212-217. IEEE, 2016.

[21] Ambrosin, Moreno, Mauro Conti, Fabio De Gaspari, and Radha Poovendran. "Lineswitch: Efficiently managing switch flow in software-defined networking while effectively tackling dos attacks." In *Proceedings of the 10th ACM Symposium on Information, Computer and Communications Security*, pp. 639-644. ACM, 2015.

[22] Wang, Tao, and Hongchang Chen. "SGuard: A lightweight SDN safe-guard architecture for DoS attacks." *China Communications* 14, no. 6 (2017): 113-125.

[23] Wang, Tao, Hongchang Chen, Guozhen Cheng, and Yulin Lu. "SDNManager: A safeguard architecture for SDN DoS attacks based

on bandwidth prediction." *Security and Communication Networks* 2018 (2018).

[24] Kaur, Karamjeet, Japinder Singh, and Navtej Singh Ghumman. "Mininet as software-defined networking testing platform." In *International Conference on Communication, Computing & Systems (ICCCS)*, pp. 139-42. 2014.

In: Advanced Security Issues …
Editor: Vijey Thayananthan

ISBN: 978-1-53615-538-9

Chapter 7

INDUSTRY 4.0 AND E_LEARNING BASED ON 5G SYSTEM

Vijey Thayananthan[1] and Esther Priyadharshini[2]
[1]Department of Computer Science,
King Abdulaziz University, Jeddah, KSA
[2]School of Education and Professional Development,
University of East Anglia, Norwich, UK

ABSTRACT

Industry 4.0 is facing many challenges in order to improve electronic learning (E_Learning) facilities with emerging 5G requirements and existing technologies. Although the 5G system provides a number of user-friendly links for developing E_Learning, users in remote areas of industry 4.0 do not get quick access whenever they want. Channel capacity should have some flexibility according to the number of users because when users' number is more than the channel capacity, users in the remote areas will have a very poor link which slows down the accessing facilities. These days, cybersecurity problems based on denial of services (DoS) create the poor links congested by the large volume of unexpected packets. These problems unnecessarily block the educational links and channels for accessing E_Learning facilities and services. To prevent these problems, a theoretical model is considered as a proposed system. Further, this model not only reduces the cybersecurity problems such as cyberbullying but also increase the channel capacity using appropriate rules and cybersecurity method. This proposed system provides the necessary solutions to enhance E_Learning services and accessing facilities.

Keywords: Industry 4.0, E_Learning, cybersecurity, 5G system

INTRODUCTION

Improving E_Learning system in industry 4.0 depends on the evolving technology and users' requirements. Although other management issues influence to improve the E_Learning services and facilities, updating technology is inevitable. In the current industrial revolution, governments of the popular countries plan to execute the E_Learning facilities to those who live in remote areas. In industry 4.0, 5G systems are going to change the educational environments where services will be the same as remote and urban areas. E_Learning services will be improved with the efficient use of a 5G system and its reliable technological capacities and limitations. Despite many technological solutions, efficient accessing facilities of E_Learning services will increase the number of participants. Regarding the 5G and 5G plus systems, future services will be merging with real-time problems such as security, energy consumptions, etc. In this situation, the cyber-physical system (CPS) will help us to provide the necessary solutions and immediate actions. Instead of improving technological capabilities, security issues should be applied to protect the E_Learning services as well as the future academic, technological devices used within the industry 4.0. However, the wrong information sent as academic contents create many unnecessary confusions to those who browse and learn the real academic contents through the E_Learning services.

As shown in Figure 1, E_Learning management system (EMS) look after some selected facilities which not only improve the educational services. The motivation of the academic services not only depends on the

EMS but also secure services which provide the necessary security solutions to all components involved in the EMS.

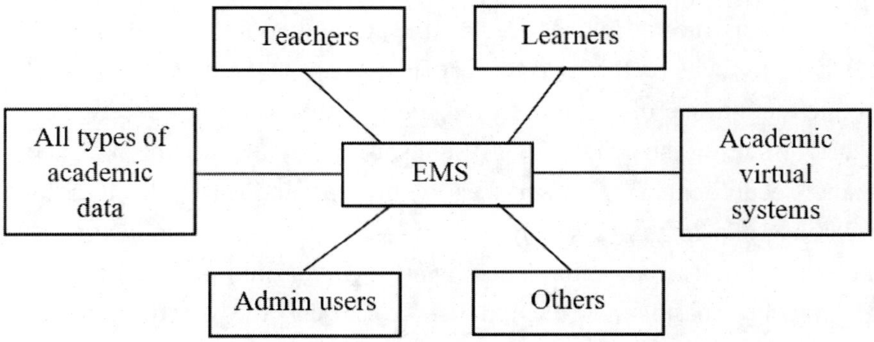

Figure 1. Components of E_Learning management system.

STATE OF THE ART

Industry 4.0

The initial focus of Industry 4.0 revolutions changed the working lifestyle through the previous knowledge obtained from hard working experienced within the old manufacturing industries. Learning through the previous mistakes, the new management of manufacturing industries started to reduce the working loads, simplify the industrial tasks, increase efficiency, etc. For instance, industry 4.0 created many applications; they are supply chain management with evolving knowledge, manufacturing according to the human working load, and secure production. With Industry 4.0, smart factories of all sectors become smart in which learning from the previous experience simplified the tasks without human supervision and reduced the overall cost.

E_Learning in Industry 4.0

The modern cities have many sections which influence government and private sectors such as IoT services of education. Researchers' point of view, they have to focus on the specific academic sector at the time to achieve the excellent solutions of the E_Learning. For instance, if we investigate the security issues of the E_Learning systems, people will enjoy the satisfaction of the secure educational services within the industry 4.0. Individual research group may take separate action which allows the industry 4.0 to improve the particular sector. There is no single definition which allows us to claim the smartness of the industry 4.0 around the world. Same time, security threats and cyber security solutions evolve with the technologies within the industry 4.0. All government and private sectors will have to work together if all citizen wants to have a secure life with a minimum cost which depends on the energy efficiency.

Advances in 5G System

Learning style on a 5G system encourages all people to understand the potential problems, precautions, etc. Despite many advances in 5G systems, E_Learning is one of the key applications, which open many books according to the problems people face during their life. Security issues based on 5G expect to provide better solutions to improve EMS, academic services and E_Learning accessing. As the latest security solution, Blockchain can be employed to prevent the cybersecurity problems which include the cyberbullying and DoS attacks.

LITERATURE REVIEW

Although the policy focused on the IoT, is a key nonfunctional requirement to every device used in the industry 4.0, authors claimed that the understanding and prediction of human behavior were important to collect and process the data for monitoring the security issues in industry 4.0. In addition, the authors provided an overview of the existing models and techniques for protecting all users and service providers within industry 4.0. Also, they considered the anonymous communication systems and limitations of policy designs in the layout of smart IoT.

Based on [1], E_Learning environments and service effectiveness are important factors for improving users' learning capacity. These factors depend on concept functionality allowed to enhance other facilities in the E_Learning. They are security, E_Learning effectiveness factors, the design process of education technology, etc. Using these factors, authors developed the theoretical framework which allows users to improve the facilities.

As discussed in [2], people use IP cameras which provide necessary physical security issues approached around the modern cities. Here, IoT plays an important role to make automation which detects the unusual objects moved or brought into the industry 4.0. Building new modern cities focus on many interconnecting components; they are IoT devices, such as sensors, actuators, smartphones and smart systems [3]. However, the Internet is to establish smart systems with a huge amount of data, termed as Big Data which is a challenging task to integrate IoT services. Here, authors present an IoT-based smart system for which next-generation Super City planning of smart cities in industry 4.0 will be considered with Big Data Analytics.

According to [4], middleware for IoT and applications in IoT environments describe the horizontal and vertical architectures. Despite the global IoT research activities used for industry 4.0 towards 2030 vision, the imminent arrival of 5G technology is revealing that current middleware approaches. According to the 5G, all middleware technology will face some improvements depended on the new 5G requirements such as bandwidth. Although authors present the concepts and architectural layers of IoT Middleware, technology helps us to develop cybersecurity solution with the IoT Middleware ability. Regarding future E_Learning, pervasive systems services are able to cope with 5G-based application requirements in IoT environments.

Smart devices are always considered with significant computational capabilities and quick transforming features to monitor and collect ambient information. Here, the authors proposed the strengths and weaknesses of various waste management models which are part of the domestic services in the industry 4.0. In their model, the framework uses smart devices and intelligent services involved with IoT-enabled waste management schemes.

According to [5], smart campuses use IoT infrastructures which need to be modernized with new designs that improve energy efficiency. Here, the authors claimed that promoting energy savings and environmental sustainability should be considered when designers are creating IoT systems from existing city infrastructures. Also, they presented energy monitoring and building management systems as an IoT infrastructure for a smart campus. As in [6], industrial cities and their traffic alert system depend on the design of IoT based traffic congestion. Here, authors claim that control unit design of industrial cities increases the service efficiency of the risk monitoring and protections. Further, a real-time traffic monitoring unit, multitasking design, sensing traffic and alert service providing unit is considered to enhance the security solutions around the industrial cities in industry 4.0.

According to [7], the authors claimed that concepts of ICT should be employed to develop the modern cities with efficient security facilities. Paper [8] demonstrates many security challenges which improve the strength of the security solutions in IoT. Modern cities in industry 4.0 need RFID readers because students and other users in the academic environments should have RFID tags which embedded the IoT devices. Here, Cryptanalysis of a novel ultra-lightweight mutual authentication protocol [9] could be useful to improve the security levels in IoT devices.

Despite the many risks, smart campuses in industry 4.0 need continuous power with low-cost and maximum cybersecurity solutions [10]. Here, accidental and purposed power failures are risks. Despite all these risks, the technique of decision modeling could be integrated with the RFID and communication networks [11].

Security of industry 4.0 technologies based on RAMI model [12] provided secure technological links for developing secure services including E_Learning activities between the remote locations.

According to [13, 14], green data storage expected to use in future educational environments, is considered with the novel cyber security solutions based on Li-Fi and quantum cryptography. According to [15-17], distance learning approaches merged with an E_Learning face not only the security problems but also speed which creates the different types cyberbullying through the IoT within the industry 4.0 where users (students and tutors) are the victims of the cyber threats.

Testbeds for E_Learning provided new features for testing and developing new educational activities [18, 19]. In industry 4.0, academic services with the E_Learning facilities need strong authentication. Based on [20], secure and robust three-factor based authentication scheme using RSA cryptosystem could be deployed to handle the authentication procedures.

According to [21-23], the safety of learners and security issues of E_Learning are considered with cyber-safety culture, and other factors such as disasters could affect the internal and extern educational environments. Here, the authors proposed a disaster recovery framework for E_Learning environment using disaster information around the affected places and inter-cloud computing. Modern cities are integrated with smart campuses [24-27] which provide smart academic environments through secure IoT frameworks. Here, examples and case studies of the modern cities and campuses provide us to establish the new smart campus within the IoT based industry 4.0 which is one of the 2030 visions. According to [28, 29], face identification and resolution framework are considered to improve security and privacy using fog computing. Again, IoT plays an important role to maintain the academic framework.

According to [30], sensors used in advanced educational environments, automation of academic services and information and Communication Technology (ICT) allow manufacturers to learn new techniques. Further, ICT and other related academic technologies provide new possibilities for lifelong learning as well as practical learning. Regarding the education based on manufacturing [31], the transition process from existing and legacy manufacturing systems to future automation and intelligent approaches will be the very expensive implementation in Industry 4.0. Despite many challenges in E_Learning, exploring gamification to support manufacturing education on industry 4.0 will be one of the innovations. The new industry 4.0 era [32] provides objectives of the engineering education rather than subjective of the engineering theory. Education and E_Learning services should encourage engineering students and learners to understand the skills of real engineering problems. In industry 4.0, curriculum and industrial perspective should have some correlations to deploy the practical and engineering skills [33]. Further, technological requirements that emerge from the industry 4.0 revolution will be the main course learning outcomes in future engineering education.

THEORETICAL MODEL

A Brief Description of the Proposed Model

With the above-mentioned literature review, a theoretical model of the IoT based device for educative environments could be designed with necessary security protocols.

Regarding secure E_Learning, reading, writing interactions, and other data handling should be protected. Thus, Figure 2 explains the details of the design including the IoT based components and security protocol, which can be used to detect and resolve potential risks of cybersecurity for E_Learning in the educative environment.

Methodology

In this research, monitoring the risks of the cybersecurity issues in E_Learning, creating suitable solutions and conducting the evaluation of the security solutions within the educational environments of the industry 4.0 are considered. The potential risks of cybersecurity for E_Learning in industry 4.0 should be detected and resolved dynamically through this model when appropriate secure IoT infrastructure is employed. Statistical performance of the reading, writing and corresponding cybersecurity policies will be monitored when students' behaviors are normal and serious, or students' concentrations are normal and poor. Using statistical performance and analysis, we can improve the monitoring of cyberbullying. Also, we can maximize the E_Learning capacities against the cybersecurity issues handled through the IoT based devices. Further, the IoT based device will monitor the exact results accurately and dynamically.

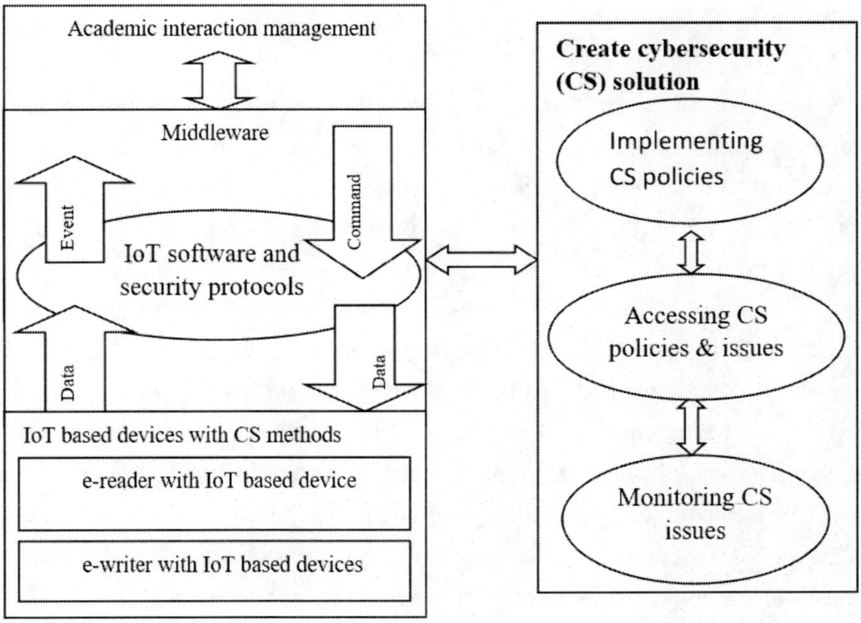

Figure 2. A theoretical model of IoT based system for the educative environment.

As shown in Figure 2, the proposed system can be designed with IoT based devices. In this design, cybersecurity policies are set according to smart campus recommendations which should be able to access during the cyber attacks when students are using the reading and writing parts of the block diagram. Further, we have formulated the efficient design of the theoretical model, secure IoT infrastructure, and security protocols. In this formulation, we employed the cybersecurity algorithm based on information security concepts. In the theoretical model, a monitoring device which collects the secured information from the different educational services helps users to improve the E_Learning capacities against the cybersecurity issues. As a suitable cyber security solution, efficient security protocols based on information security can be employed as an alternative method. Also, protocols of cyberbullying merged within the E_Learning systems could be added as a part of the methodologies.

Security in E_Learning Environments

Security in the current E_Learning system may not be enough because interactions and academic services are increasing within the E_Learning system and the technologies. So, in our approach, security can be increased according to the request which is the level of the protection for particular service used in the academic environments.

The design of academic interaction management is important issues in the In this design, the following management approaches are identified to improve the educational environments within the industry 4.0.

Campus Management: The cybersecurity issues in campus management need some strong policies which protect the academic services assigned in educative environments. Handling large data in academic services is still not 100% secured because cybersecurity solutions are not efficient. Despite the successful campus management, implementing cybersecurity policies and updating these policies based on the evolving technologies and cyber threats are not easy for managing 100% security in the campus management. Therefore, by introducing IoT based devices integrated with security in the campus management system, all these flaws can be solved perfectly. Here, the level of security should be considered according to the applications.

Library Management: Accessing online approach in the library is very slow during the busy or peak time because many users utilize the services according to their priorities. Despite many accessing policies, library management needs to update the accessing cybersecurity policies according to the time, priorities, etc. Further, library management supposes to provide secure E_Learning facilities which maintain the library system, data handling, storage of e_books and administrations. Within the library environment, the IoT based devices can be considered to improve the secure E_Learning systems in the educational environments.

Employee's access and management: The education environment should maintain the monitoring devices which provide all employees to access and manage their confidential information. Although many cybersecurity policies are available, the following two approaches improve the employees' access and management system with security:

1. Managing passwords in a security-rich fashion.
2. Integrating with an enterprise identity management system.

In our theoretical model, we introduced the cybersecurity policies in three different ways which allow us to analyze the improvement of the E_Learning capacities. Firstly, implementing cybersecurity policies in the proposed model should be dynamic. Secondly, accessing online systems and cybersecurity policies for accessing should have maximum security levels. Finally, monitoring facilities involved in the educational environments depend on the privacy issues of the cybersecurity policies. Managing passwords with all possible cybersecurity policies are one of the important tasks in employees' access and management schemes.

EXPECTED RESULTS

According to the theoretical model, educational environments of industry 4.0 need some inaugurations. Thus, we can compare the expected improvements with the conventional and proposed approach. As shown in Figure 3, results show three different cases which are currently facing difficult situations during the cyberattacks depended on the potential risks. Implementing updated cybersecurity policies regularly allow us to improve the security issues of E_Learning about 10%. Same way, cybersecurity policies for accessing increase the accessing time of E_Learning services by 15%. Monitoring all cybersecurity issues including all possible cyber attacks increase the overall E_Learning facilities in educational environments by 20%. Academic services managed between the educational environments and users need to be secured using appropriate algorithms used in cyber security solutions. Here, results are compared between the traditional and IoT based system when our proposed model measures the potential risks of cybersecurity for E_Learning in the selected educational environment of the industry 4.0. Although this theoretical model should be able to detect all the cyber attacks and threats dynamically, results will be depending on many factors such as cybersecurity policies of industry 4.0.

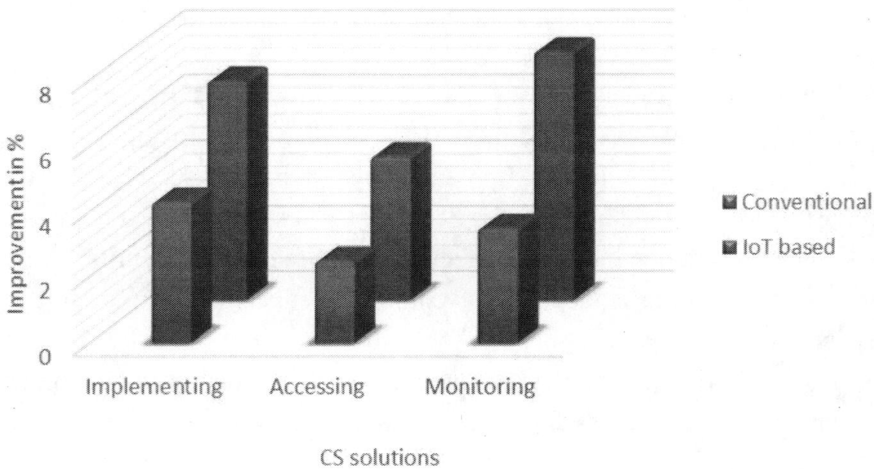

Figure 3. Expected improvements when CS is employed within the E_Learning.

Learning in industry 4.0 takes us to involve many activities and services we face in our life. Although 5G provides technological improvements to most of the services, users expectations and knowledge create new applications through the E_Learning. In this chapter, the cybersecurity issue is one of the problems, which prevent accessing services and other applications.

DISCUSSION

Solutions of Cyberbullying in Industry 4.0

Industry 4.0 provides an improvement in E_Learning solutions based on evolving technology requirements. Despite many solutions, interruptions and DoS attacks are still within the system which might be improved through the 5G plus technology. Cyberbullying is one of the recent attacks which damage not only the personal or targeted user but also EMS. All E_Learning services depend on the data which are in the text or audio or video format. Thus, data traffic should be managed securely because cyberbullying employes one of the data formats to block the accessing channels and services provided by EMS.

Accessing E_Learning in 5G Environments

Although accessing E_Learning contents represented by specific formats such as text or audio or video or combination of all is available in 5G environments, attacks related to cyberbullying prevent the accessing facilities. In the future, E_Learning facilitated through the virtual system will provide the online teaching method. Using the virtual system, students can share the teaching slides while the teacher is presenting/demonstrating/ teaching in the class.

CONCLUSION

We have studied the necessary issues of potential risks and cybersecurity problems for improving E_Learning facilities. In these studies, finding potential cyber risks, a theoretical model for detecting cyber attacks and secure educational environment with E_Learning facilities in industry 4.0 are considered. In the theoretical model, expected solutions for protecting academic services based on IoT based 5G infrastructure are focused. Using a theoretical model, we can maximize E_Learning services and channel capacities through the best security policies. During virtual learning, accessing teaching materials depends on many factors such as security, evolving technologies, etc. In future work, adding Blockchain technology into the proposed theoretical model will allow us to improve the security solutions in all environmental conditions.

REFERENCES

[1] Haq, Anwar ul, George Magoulas, Arshad Jamal, Asim Majeed, and Diane Sloan. "Users' perceptions of e-learning environments and services effectiveness: The emergence of the concept functionality model." *Journal of Enterprise Information Management* 31, no. 1 (2018): 89-111.

[2] García, Cristian González, Daniel Meana-Llorián, B. Cristina Pelayo G-Bustelo, Juan Manuel Cueva Lovelle, and Nestor Garcia-Fernandez. "Midgar: Detection of people through computer vision in the Internet of Things scenarios to improve the security in Smart Cities, Smart Towns, and Smart Homes." *Future Generation Computer Systems* 76 (2017): 301-313.

[3] Rathore, M. Mazhar, Anand Paul, Awais Ahmad, and Gwanggil Jeon. "IoT-based big data: From smart city towards next generation super city planning." *International Journal on Semantic Web and Information Systems (IJSWIS)* 13, no. 1 (2017): 28-47.

[4] Amaral, Leonardo Albernaz, Everton de Matos, Ramão Tiago Tiburski, Fabiano Hessel, Willian Tessaro Lunardi, and Sabrina Marczak. "Middleware Technology for IoT Systems: Challenges and Perspectives toward 5G." In the *Internet of Things (IoT) in 5G Mobile Technologies*, pp. 333-367. Springer, Cham, 2016.

[5] Bates, Oliver Emile Glaves, and Adrian John Friday. "Beyond data in the smart city: learning from a case study of re-purposing existing campus IoT." *IEEE Pervasive Computing* 16, no. 2 (2017): 54-60.

[6] Li, Shancang, Li Da Xu, and Shanshan Zhao. "5G Internet of Things: A survey." Journal of Industrial Information Integration (2018).

[7] Stolwijk, C. C. M., and L. M. Punter. *ICT developments in smart industry field labs*. No. TNO 2017 R10006. TNO, 2017.

[8] Samaila, Musa G., Miguel Neto, Diogo AB Fernandes, Mário M. Freire, and Pedro RM Inácio. "Security challenges of the Internet of Things." *Beyond the Internet of Things*, pp. 53-82. Springer, Cham, 2017.

[9] Tewari, Aakanksha, and B. B. Gupta. "Cryptanalysis of a novel ultra-lightweight mutual authentication protocol for IoT devices using RFID tags." *The Journal of Supercomputing* 73, no. 3 (2017): 1085-1102.

[10] Agiwal, Mamta, Abhishek Roy, and Navrati Saxena. "Next-generation 5G wireless networks: A comprehensive survey." *IEEE Communications Surveys & Tutorials* 18, no. 3 (2016): 1617-1655.

[11] Thayananthan, Vijey, Ahmed Alzahrani, and Muhammad Shuaib Qureshi. "Efficient techniques of key management and quantum cryptography in RFID networks." *Security and Communication Networks* 8, no. 4 (2015): 589-597.

[12] Flatt, Holger, Sebastian Schriegel, Jürgen Jasperneite, Henning Trsek, and Heiko Adamczyk. "Analysis of the Cyber-Security of industry 4.0 technologies based on RAMI 4.0 and identification of requirements." In *Emerging Technologies and Factory Automation (ETFA), 2016 IEEE 21st International Conference on*, pp. 1-4. IEEE, 2016.

[13] Shaikh, Riaz Ahmed, and Vijey Thayananthan. "Hop-by-hop trust evaluation algorithm for identity anonymous sensor networks." *IETE Journal of Research* 61, no. 2 (2015): 154-159.

[14] Thayananthan, Vijey, and Aiiad Albeshri. "Big data security issues based on quantum cryptography and privacy with authentication for the mobile data center." *Procedia Computer Science* 50 (2015): 149-156.

[15] Goumopoulos, Christos, Petros Nicopolitidis, Damianos Gavalas, and Achilles Kameas. "A distance learning curriculum on pervasive computing." *International Journal of Continuing Engineering Education and Life Long Learning* 27, no. 1-2 (2017): 122-146.

[16] Cho, Sok Pal, and Jeom-Goo Kim. “E-Learning Based on Internet of Things.” *Advanced Science Letters* 22, no. 11 (2016): 3294-3298.

[17] Gabor, Andrei Marius, Marius Constantin Popescu, and Antoanela Naaji. “Security Issues Related To E-Learning Education.” *International Journal of Computer Science and Network Security (IJCSNS)* 17, no. 1 (2017): 60.

[18] Oda, Tetsuya, Keita Matsuo, Leonard Barolli, Masafumi Yamada, and Yi Liu. “Design and implementation of an IoT-based e-learning testbed.” *International Journal of Web and Grid Services* 13, no. 2 (2017): 228-241.

[19] Yamada, Masafumi, Miralda Cuka, Yi Liu, Tetsuya Oda, Keita Matsuo, and Leonard Barolli. “Evaluation of an IoT-based e-learning testbed: Performance of OLSR protocol in a NLoS environment and mean-shift clustering approach considering electroencephalogram data.” *International Journal of Web Information Systems* 13, no. 1 (2017): 2-13.

[20] Ali, Rifaqat, and Arup Kumar Pal. “A secure and robust three-factor based authentication scheme using RSA cryptosystem.” *International Journal of Business Data Communications and Networking (IJBDCN)* 13, no. 1 (2017): 74-84.

[21] Kritzinger, Elmarie. “Cultivating a cyber-safety culture among school learners in South Africa.” *Africa Education Review* 14, no. 1 (2017): 22-41.

[22] Majeed, Asim, Said Baadel, and Anwar Ul Haq. “Global Triumph or Exploitation of Security and Privacy Concerns in E-Learning Systems.” In *International Conference on Global Security, Safety, and Sustainability*, pp. 351-363. Springer, Cham, 2017.

[23] Togawa, Satoshi, and Kazuhide Kanenishi. “Building a Disaster Recovery Framework for e-Learning Environment Using Disaster Information and Inter-cloud Computing.” In *Advances in Human Factors, Business Management, Training, and Education*, pp. 901-908. Springer, Cham, 2017.

[24] Pinto, Luciana Gomes Pereira, Regiane Relva Romano, and Matheus Akira Tomoto. "From the University to Smart Cities—How Engineers Can Construct Better Cities in BRIC's Countries: A Real Case from Smart Campus FACENS." In *Advances in the Human Side of Service Engineering*, pp. 347-354. Springer, Cham, 2017.

[25] Rathore, M. Mazhar, Awais Ahmad, and Anand Paul. "IoT-based smart city development using big data analytical approach." In *Automatica (ICA-ACCA), IEEE International Conference on*, pp. 1-8. IEEE, 2016.

[26] Sari, Marti Widya, Prahenusa Wahyu Ciptadi, and R. Hafid Hardyanto. "Study of Smart Campus Development Using Internet of Things Technology." In *IOP Conference Series: Materials Science and Engineering*, vol. 190, no. 1, p. 012032. IOP Publishing, 2017.

[27] Khatoun, Rida, and Sherali Zeadally. "Cybersecurity and privacy solutions in smart cities." *IEEE Communications Magazine* 55, no. 3 (2017): 51-59.

[28] Hu, Pengfei, Huansheng Ning, Tie Qiu, Houbing Song, Yanna Wang, and Xuanxia Yao. "Security and privacy preservation scheme of face identification and resolution framework using fog computing in the internet of things." *IEEE Internet of Things Journal* 4, no. 5 (2017): 1143-1155.

[29] Rayes, Ammar, and Samer Salam. *Internet of Things from hype to reality: The road to digitization*. Springer, 2016.

[30] Tvenge, Nina, and Kristian Martinsen. "Integration of digital learning in industry 4.0." *Procedia Manufacturing* 23 (2018): 261-266.

[31] Paravizo, Esdras, Omar Cheidde Chaim, Daniel Braatz, Bernd Muschard, and Henrique Rozenfeld. "Exploring gamification to support manufacturing education on industry 4.0 as an enabler for innovation and sustainability." *Procedia Manufacturing* 21 (2018): 438-445.

[32] Onar, Sezi Cevik, Alp Ustundag, Çigdem Kadaifci, and Basar Oztaysi. "The Changing Role of Engineering Education in Industry 4.0 Era." In *Industry 4.0: Managing the Digital Transformation*, pp. 137-151. Springer, Cham, 2018.

[33] Hernández-Muñoz, Guadalupe Maribel, Lizbeth Habib-Mireles, Flor Araceli García-Castillo, and Fernando Montemayor-Ibarra. "Industry 4.0 and Engineering Education: An Analysis of Nine Technological Pillars Inclusion in Higher Educational Curriculum." In *Best Practices in Manufacturing Processes*, pp. 525-543. Springer, Cham, 2019.

In: Advanced Security Issues …
Editor: Vijey Thayananthan
ISBN: 978-1-53615-538-9

Chapter 8

Traffic Load Balancing Based on Routing Algorithms for Content-Centric Networking

***Muhammad Shuaib Qureshi*[1], *Muhammad Bilal Qureshi*[2] *and Hameed Hussain*[3]**
[1]Department of Computer Science, King Abdulaziz University, Jeddah, KSA
[2]Department of Computer Science, Shaheed Zulfikar Ali Bhutto Institute of Science & Technology, Islamabad, Pakistan
[3]Department of Computer Science, University of Buner, KPK, Pakistan

Abstract

In Content-Centric Networking (CCN), video streaming and content sharing increase the growth rate of mobile traffic. The Internet supports services to mobile devices such as smart phone and ipads with many challenges which may include balancing traffic problems etc. Mobility support for CCN faces massive content deliveries through the current Internet architecture. The traffic analysis can improve the performance of CCN in which we have chosen the minimum average distance and critical generation rate as basic traffic metrics. In this analysis, traffic load is balanced or uniformly distributed using the aforementioned two traffic metrics. For the efficient and reliable CCN, the minimum average distance used for the content communication between the nodes should be short. These problems are manageable with analyzing traffic load balancing that depends on the employed routing algorithms. In this research, the Shortest Path routing algorithm is used as a method to analyze the traffic load within the CCN. As a result, the shortest path selection improves the traffic load and provides efficient and reliable CCN. Finally, a necessary comparison of the traffic in CNN shows the supremacy of the proposed approach over the existing state-of-the-art counterparts.

Keywords: content-centric networking, load balancing, routing algorithm, minimum average distance, critical generation rate

INTRODUCTION

The rapid use of contents sharing, delivering and transmitting increases the traffic congestions in many communication networks including CCN. The traffic load depends on the two basic messages; Interest and Data established as communication in CCN. The *Interest* is the data message requested by the clients or users. The Data message not only contains the name but also cryptographic techniques. Traffic in each CCN node varies because traversing data packets and a buffer memory of the cached node have many changes. These changes contain the details of the packets, the status of the memory, operation and environmental conditions, and so on. The different online sources define Interest and Data as below.

Interest: A consumer puts the name of a desired piece of data into an Interest packet and sends it to the network. Routers use this name to forward the Interest towards data producer(s).

Data: Once the Interest reaches a node that has the requested data, the node will return a Data packet that contains both the data name and content together with a signature by the producer's key which binds the two. This Data packet follows in reverse the path taken by the Interest to get back to the requesting consumer.

As the traffic volume increases, the traditional content delivery approaches take complicated paths through the heavy traffic load, but CCN is considered as a fast and efficient approach because it uses the content name. Analyzing traffic load in CCN, the content deliveries can be improved by using appropriate routing techniques which depends on the traffic metrics. The appropriate load balancing and routing algorithms can control the traffic metrics such as minimum average distance and critical generation rate. The routing algorithms control while load balancing optimizes the network traffic problems. In this research, we have introduced the new technique that employs the appropriate routing and load balancing algorithms which solve traffic problems in CCN.

In 2013, about 4 billion people used a mobile phone, but the volume of the traffic was reasonable with the features set up by the users. Currently, the phone has not only more features but also improved services with the latest CCN technology. As the number of users increases, the management of high volume traffic becomes challenging. This mess needs not only to reduce the traffic overhead but to improve memory usage too controlled by the traffic metrics. The caches employed in the CCN can substantially reduce the traffic on the links used in applying CCN.

Most of the traffic flow through Internet content such as web pages and videos. Nowadays, people access the Internet through mobile devices with many communication technologies. Further, machine to machine, human to machine and machine to human communication also depends on the Internet connections that increase the traffic volume.

THE CCN AND TRAFFIC MODELING

Although the traffic background of the CCN is important, we focus on some features that are most important to analyze the traffic control in CCN. The history of CCN and background knowledge can be found in the existing literature. Traffic characterization through modeling and analysis provides necessary predictions of traffic within the CCN environments. Traffic characterization is the stochastic nature of any network traffic which encompasses Internet and multimedia traffic. Traffic load depends on many factors such as basic network operating parameters, environmental conditions, and human behaviors, etc. Further, fast convergence is also an important parameter in CCN traffic modeling depends on the Shortest Path (SP) routing.

Basic Features for Traffic Control

There are some basic approaches that control the traffic based on Interest and Data packets in the CCN. One of such approaches is the naming approach of CCN packet that holds the details of the CCN name with the notion of a content name. In traffic control, specifying bound for the payload is very important that holds several kilobytes of CCN packet. Another feature is forwarding the data which can change the data traffic between the nodes used in the CCN. Sensitive traffic such as interactive audio or video contents should be monitored during the CCN packet transmission that might cause some delays between the source and destination nodes. Traffic modeling may have different levels such as peak and off-peak in which CCN traffic depends on the users' total usage time.

Traffic can flow for exchange between two different CCNs. If the traffic is heavy and many CCN packets are waiting to be transmitted, we need to reduce the time delay such as queuing time. The time delay depends on the transmission rate of the network link used in the CCN and the nature of the arriving traffic which may be periodic or burst. Traffic Intensity given in Eq. 1 plays a vital role in addressing traffic problems in CCN

$$Intensity = \frac{La}{R} \tag{1}$$

where a is the unit of the packets per second, R is the transmission rate in bits per second (bps), and L is the packet size in bits.

Traffic Modeling

Traffic model encompasses basic traffic metrics such as throughput and round trip time, but the proposed traffic model uses path distance and packet rate generation metrics. We develop a CCN traffic control framework based on the distance of the shortest path and the critical rate of the packet generation. Regarding traffic modeling, the basic traffic model that transmits the CCN packets between source and destination uses bandwidth sharing and router forwarding strategies. When we compare the existing network based on IP address, CCN has some important properties that help us to improve the traffic loads.

Traffic Load Balancing

The amount of content, feature, and service increases traffic load in CCN and other networks as well. In order to improve traffic problems, it is wised to distribute CCN traffic using load balancing concepts. Load balancing solves the above-mentioned issues by reducing congestion and averaging workload among nodes. There are many types of load balancing algorithms available, and a lot of research has been conducted in the area over the years. Two types of load balancing algorithms were tested in the evaluation of the distributed system to determine which type of algorithm can produce improved throughput. Load balancing can be used to increase the access rate that will lower the traffic intensity on the access link used in the CCN. Also, proper load balancing approach can result in negligible delays between the routers. The traffic load can also be distributed using multiple DNS servers and caches that reduce the query traffic received as Interest packets used in the CCN.

Traffic Routing

If there is no place to store the CCN packets in the content memory (buffer), the router will drop the packets. It depends on the router traffic intensity used in the CCN. Dynamic routing algorithm changes the routing path depends on the CCN traffic load that may be Data or Interest packets.

RELATED WORK

Li et al. [1], analyzed the network routing traffic performance using holistic network model for CCNs. They studied the performance and optimization of routing contents from source to destination nodes. Uses of contents are continuously increasing with the population. It means that traffic load increases within the CCN which include a mobile social network, 5G, etc. The traffic load can be distributed to improve the content-centric mobile network in 5G [2]. Network traffic and response delay can be reduced when Interested contents are cached in the content store [3].

As mentioned by Carofiglio et al., in [4], the request arrival process is modeled through a Markov Modulated Rate Process (MMRP). Generally, MMRP is the modified version of the Poisson theory used in traffic modeling, but the authors have studied and analyzed these two approaches for CCN. Finally, they employed the MMRP approach in the context of chunk-based CCN which helped to develop their proposed model. Rossini et al. [5], showed that network efficiency depends on the multipath forwarding capabilities. In the current research, multipath forwarding and caching policies are playing an important role in analyzing the performance of CCN. The caching policies improve network efficiency thereby reducing traffic problems and computational overhead.

Regarding mobility management, the authors in [6-7], focused on faster service and appropriate routing in CCN environment For improving the faster service access latency, they analyzed the user created mobile contents instead of other basic features. The named CCN is a better approach than the IP address because the later is limited by the network technology. Based on the discussed routing and mobility management concept, we have employed the load balancing technique that improves the access latency. The normal routing provides connections which establish a route in CCN but handover procedure for which real-time communication between the nodes is not possible [8]. CCN does not support any handover procedures for mobile source nodes, however, different approaches can help in minimizing delay and traffic congestion. The energy efficient caching for CCN architecture ensure that all nodes used in the CCN are active to exchange the packets and necessary data with minimum delay without any further waiting [9]. The saved energy not only helps to reduce the network traffic but also improves user access. Furthermore, the application may be extended to develop a content-centric Internet architecture [10].

Based on previous findings, Asaeda et al. [11], implemented a measurement tool on a CCN router. Using this tool, the CCN traffic influences the forwarding path that carries the information per name prefix. It is also investigated that traffic dominated in CCN depends on the cache information, content popularity, and cache capacity. Regarding the traffic load analysis on information-centric networking, details of caching functionalities, performance and other relevant points are discussed in [12]. Wang et al. [19], discussed load-aware potential-based routing for improving user QoS which depends on the CCN. Despite the models developed in the previous research projects, the authors in [19] addressed the potential demands and designed a new model for the edge communication of smart grid with CCN.

According to Cai et al. [20], CCN provides an alternative step to solve some potential problems such as the host-centric Internet. Choosing the optimal forwarding interface for analyzing the Interest packet and selecting an optimized routing path are some of the current challenges that need to address. To improve the problem, authors propose a Software Defined Status Aware Routing (SD-STAR) mechanism which not only controls the packet delay but also node load in the controller used in the CCN.

Information-Centric Network (ICN) will be one of the merging networks in the future Internet [21]. Further, it will encourage end-customers to handle efficient data services through the CCN. In ICN, routing and forwarding strategy may be better designed for transmitting data dynamically.

According to Li et al. [23], efficient content delivery and retrieval using CCN are the most important features in the 5G network. The challenges of IoT based 5G will have some influences on Name Data Networking (NDN) and network coding [24]. Using network coding with NDN improves the CCN where throughput and efficiency increase with network and content delivery respectively. Combination of Edge-Centric Computing (ECC) and CCN provides the basis for ECC architecture and merges with the 5G network [25]. The integration of Software-Defined Networking (SDN) and ECC enhance manageable and flexible services.

PROPOSED RESEARCH

In the proposed model, we have combined the Shortest Path (SP) routing and Load Balancing (LB) algorithms for efficient utilization and improvement in the future traffic load in CCN. To activate the proposed research, the following techniques need to be addressed with routing and load balancing theories.

Theoretical Model

This model contains routing processing that controls the CCN traffic and loads balancing to keep the CCN traffic in a steady state. It also models the CCN traffic demand between the source and destination points where workload moves over time. Regarding the routing design, many researchers have already compared the IP and CCN node models. Routing optimization based on load balancing has already been moved within which wireless network technology is available. In our theoretical model, routing optimization improves the efficiency of data transmission without occurring any congestion.

Figure 1 shows the CCN traffic model in which Interest and Data packets affect the traffic during transmission, forwarding, etc. For efficient and reliable transmission and forwarding, the traffic load is distributed using load balancing concepts. This mechanism minimizes the traffic load within the CCN when Data and Interest packets are in transit. By applying hybridization of routing and load balancing algorithms, traffic in CCN improves. An Access Point (AP) deals with the monitoring, handling and accessing data traffic of the contents between two or more networks involved in the CCN.

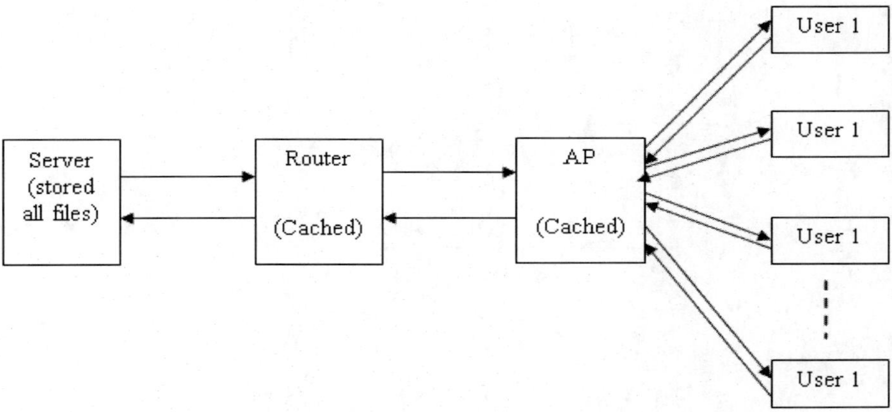

Figure 1. CCN model with traffic load.

Proposed Model Based on Traffic Load Balancing

Load balancing in complex CCN traffic plays an important role that not only balances the traffic load but also optimizes the network routing. In this model, the CCN, traffic control framework, based on the minimum average distance and generation rate of each node provides better data transmission. Average minimum distance and generation rate of each node in the CCN depends on Shortest Path (SP) routing and Simulated Anealing (SA) algorithms. Here, the average number of generated packets means CCN packets in each node during traffic flow.

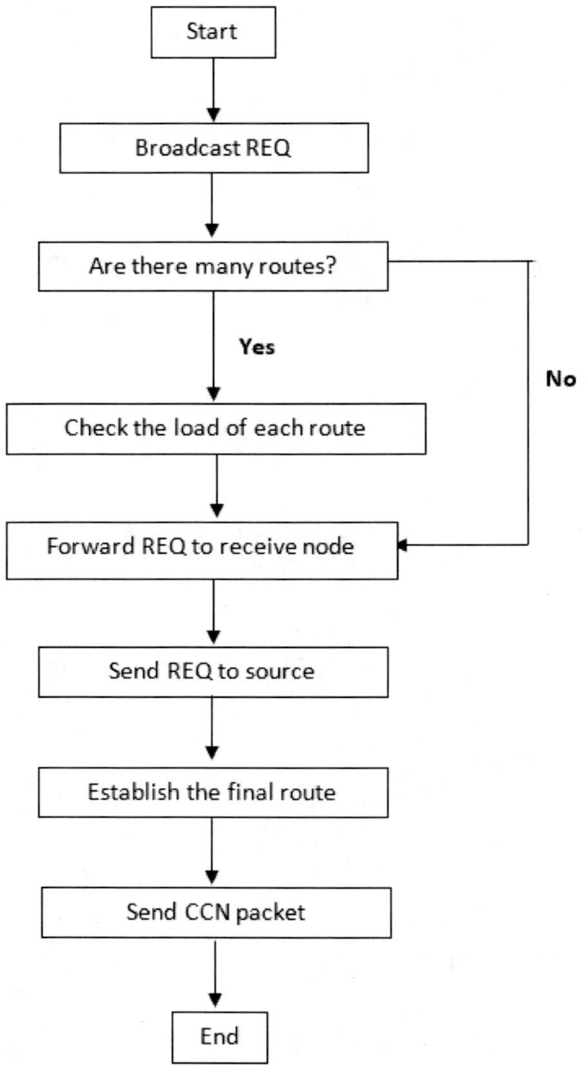

Figure 2. Traffic load balancing for CCN.

As shown in Figure 2, the CCN takes the best path through the traffic load balancing that depends on routing algorithms. In this approach, routing is the dedicated process of finding the best available path among the available CCN nodes that hold the CCN packets for transmission or forwarding. The Route Request (RREQ) packet contains Routing Cost (RC), which represents the traffic load of each intermediate node between source and destination. This field is updated upon moving the packet between intermediate nodes, and then the destination node collects RC information to select a path with minimum RC. The CCN nodes forward routing requests to gateways; the gateways can play a critical role in load balancing by grouping the nodes in multiple non-overlapping groups. The new node will broadcast RREQ where first gateway receiving this request will assign a group identifier to the node to be a member of the group. It sends Route Response (RREP) packet with the group identifier back to the source node. If the intermediate node is a member of the group and receives this RREQ, it will allow the new node to join the group and send back RREP packet.

Implementation

The proposed model is set up with the same SP routing procedures and conditions suggested by the Wu et al. [13], in addition to the load balancing mechanism which provides an extra improvement in the traffic analysis.

Although the procedure contains a lot of mathematical concepts, yet some necessary formulations are considered in the subsequent sections. The SP routing algorithm is the best mechanism for traffic analysis when the distance with the network size is considered. It is well known and widely used routing algorithm for improving network transmission efficiency. The node usage probability characterizes the traffic problem, and SA algorithm controls the traffic using minimum average distance and node usage. The SP routing strategy for load balancing helps in reducing high traffic loads at some devices in the CCN, unnecessary traffic congestion, and delay of the whole CCN environment.

The average minimum distance depends on the SP routing algorithm, but it does not support the generation rate.

The SA technique is implemented to calculate the critical generation rate which characterizes the traffic load.

A new approach that uses load balancing and routing strategies for improving traffic performance depend on the average minimum distance and critical generation rate.

PERFORMANCE OF THE PROPOSED RESEARCH MODEL

The combination of routing algorithms with load balancing technique improves performance when traffic loads are uncontrollable in CCN. For this purpose, we analyze the following points employed in the theoretical model. Firstly, the average minimum distance determines the shortest path identified by the SP routing algorithm. Secondly, the generation rate was calculated using the SA algorithm as new traffic that is controllable.

The Average Distance for Traffic Analysis

As shown in Figure 3, the minimum average distance for the proposed and existing models is compared. As traffic metric, the average distance is the distance which the proposed model selects as the shortest path based on the specific routing algorithm. As a traffic load, the total number of paths passed through a particular node in the CCN is calculated. Using these steps, the minimum average distance D is the ratio of the total number of paths and square of the total number of nodes used in the CCN as shown in Eq 2.

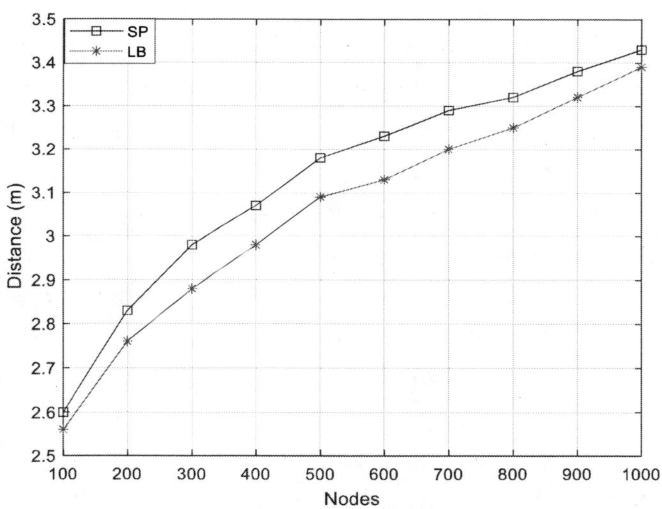

Figure 3. Average minimum distance for the traffic analysis.

$$D = \frac{C}{N^2} \tag{2}$$

Where, C is the total number of paths in a particular node, and N is the total number of nodes.

Critical Generation Rate for the Traffic Analysis

Critical generation rate was calculated using the SA algorithm as a new traffic measurement which is controllable with selected parameters given in Eq 3. Generation rate G is inversely proportional to the average minimum distance D and node usage probability U.

$$G = \frac{K}{DU} \tag{3}$$

Here, the constant K where $K = \frac{R}{N}$ depends on the total number of nodes N in the CCN and the transmission capacity R of the node respectively.

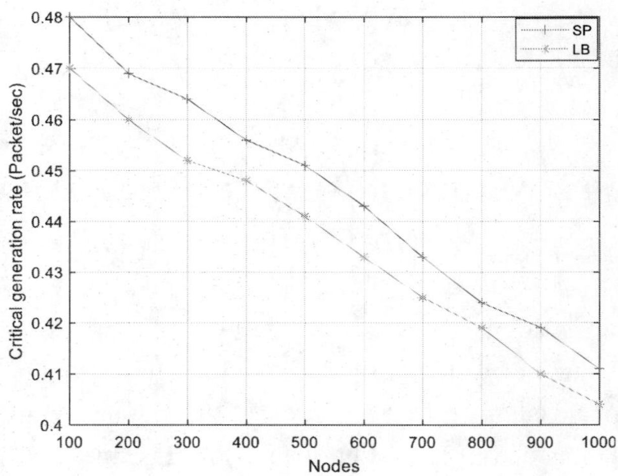

Figure 4. The critical generation rate for traffic analysis.

As shown in Figure 4, it can be seen that the critical generation rate based on the proposed approach performs better. In this analysis, average CCN packet transmission performs well because critical generation rate reduces the traffic.

Traffic Analysis with CCN Capacity

When the network or CCN capacity is increased, it can transmit more packets without any traffic problems. In this research, traffic analysis depends on the average minimum distance and generation rate. The results obtained in Figures 3 and 4 show that these metrics certainly increase the CCN capacity.

It is necessary to address which performance analysis can be proved effective in the proposed research. Although the most important traffic metrics for the CCN are the throughput and round trip time *RTT* between the source node and the destination node. In the proposed model, the average distance is considered. In CCN routers, load balancing plays an important role in identifying the shortest path that helps in analyzing network traffic. Although caches replace traditional buffers to improve traffic problems, the proposed approach enhances users' access. So, considering all these performance metrics, future CCN will use values of D and G which help in managing the traffic load.

Improvement in Traffic Management in the Future CCN

Efficient traffic management in CCN is valid for many reasons. It not merely provides heavy traffic cost reductions but also improves traffic optimization and energy efficiency. Most of the routing algorithms improve traffic on the network only, but load balancing mechanism optimizes network traffic. Using both load balancing and routing techniques as a hybrid technique, the CCN performance can be improved with minimum traffic. This hybridised technique can efficiently adjust and distributes traffic load according to the traffic congestions and delays.

In many ways, routing works better in CCN than IP and provides some advantages; they are a multi-destination, converging, etc. This advantage certainly improves traffic performance and allows to manage it efficiently. In the CCN traffic analysis, the length of the CCN names is kept reasonable because in most cases, CCN names are longer than the IP identifier. Theoretically, the CCN works well and finds the shortest path easily without any topology that reduces the use of bandwidth when Interest packet is in transit.

Name types, routing table size, etc. are the few examples of scalability used in CCN where we need specific routing algorithms to control traffic load. According to [14], authors introduced the user traffic modeling that was useful to improve the new services, network planning framework and large-scale network optimization. In this traffic model, the user-centric approach is introduced because user behavior patterns depending on the contents. The CCN uses a pull-based mechanism to transfer the data which then exchange the traffic load between the nodes [15]. The authors have introduced three functions to achieve scalability on large-scale push applications. In this work, traffic, and scalability used in different applications evaluated as performance metrics of the CCN.

Another dimension of CCN management problem is to find a solution for managing a massive amount of mobile traffic for delivering high-quality video contents through the current network infrastructure. The previous work [16] addressed this problem using content-centric mobile networks (CCMNs) approach, but still leaves holes for further improvements. The incorporation of cache mechanism in CCMN technique [17] tremendously improved the performance yet, potential energy parameter [22] produced enhanced efficiency in CCMN technique in terms of content diversity, redundancy, users, request hop, and so on.

The aforementioned improvements enhanced the capabilities of CCNs, but constant flow of web contents from outside ISP creates many unexpected problems such as infrastructure cost, security, etc. which reduce the overall profit [18]. The above-discussed problems can be solved by using micro

CDNs in ISP access and back haul networks. This enhanced technology will not only reduce the overall cost of the web traffic within the ISP infrastructure but also help in improving the user QoS.

Conclusion and Future Work

According to traffic modeling in this research, CCN traffic is the superposition of the flows. We have studied and analyzed the CCN traffic control that depends on the basic features of CCN. Finally, we have analyzed the theoretical model based on routing algorithms and load balancing for controlling traffic in CCN. According to the early results, the proposed approach gives better performance. It means that we can control the traffic load using shortest distance and critical generation rate. Some features and payload details are very important factors used in the theoretical model.

In future work, traffic analysis will be possible with energy efficient routing and load balancing algorithms. Also, this research leads to investigate virtual buffer and bufferless managements at the router which can further improve traffic load in CCN.

References

[1] Li, Yanhua, Haiyong Xie, Yonggang Wen, Chi-Yin Chow, and Zhi-Li Zhang. "How much to coordinate? Optimizing in-network caching in content-centric networks." *IEEE Transactions on Network and Service Management* 12, no. 3 (2015): 420-434.

[2] Su, Zhou, and Qichao Xu. "Content distribution over content-centric mobile social networks in 5G." *IEEE Communications Magazine* 53, no. 6 (2015): 66-72.

[3] Chu, Weibo, Lifang Wang, Haiyong Xie, Zhi-Li Zhang, and Zejun Jiang. "Network delay guarantee for differentiated services in content-centric networking." *Computer Communications* 76 (2016): 54-66.

[4] Carofiglio, Giovanna, Massimo Gallo, Luca Muscariello, and Diego Perino. "Modeling data transfer in content-centric networking." In *Proceedings of the 23rd International Teletraffic Congress*, pp. 111-118. International Teletraffic Congress, 2011.

[5] Rossini, Giuseppe, and Dario Rossi. "Evaluating CCN multi-path interest forwarding strategies." *Computer Communications* 36, no. 7 (2013): 771-778.

[6] Lee, Jihoon, DaeYoub Kim, Myeongwuk Jang, and Byoung-joon Lee. "Mobility management for mobile consumer devices in content-centric networking (CCN)." In *Consumer Electronics (ICCE), 2012 IEEE International Conference on*, pp. 502-503. IEEE, 2012.

[7] Lee, Jihoon, Sungrae Cho, and Daeyoub Kim. "Device mobility management in content-centric networking." *IEEE Communications Magazine* 50, no. 12 (2012).

[8] Luo, Yunqi, Jonas Eymann, and Andreas Timm-Giel. "Mobility support for content centric networking." *Telecommunication Systems* 59, no. 2 (2015): 271-288.

[9] Li, Jun, Bin Liu, and Hao Wu. "Energy-efficient in-network caching for content-centric networking." *IEEE Communications Letters* 17, no. 4 (2013): 797-800.

[10] Zahariadis, Theodore B., Petros Daras, Jan Bouwen, Norbert Niebert, David Griffin, Federico Álvarez, and Gonzalo Camarillo. "Towards a Content-Centric Internet." In *Future Internet Assembly*, pp. 227-236. 2010.

[11] Asaeda, Hitoshi, Kazuhisa Matsuzono, and Thierry Turletti. "Contrace: a tool for measuring and tracing content-centric networks." *IEEE Communications Magazine* 53, no. 3 (2015): 182-188.

[12] Hail, Mohamed Ahmed M., Marica Amadeo, Antonella Molinaro, and Stefan Fischer. "On the performance of caching and forwarding in information-centric networking for the IoT." In *International Conference on Wired/Wireless Internet Communication*, pp. 313-326. Springer, Cham, 2015.

[13] Wu, Jiajing, K. Tse Chi, and Francis CM Lau. "Optimizing performance of communication networks: An application of network

science." *IEEE Transactions on Circuits and Systems II: Express Briefs* 62, no. 1 (2015): 95-99.

[14] Svigelj, Ales, Radovan Sernec, and Kemal Alic. "Network traffic modeling for load prediction: a user-centric approach." *IEEE Network* 29, no. 4 (2015): 88-96.

[15] Virgilio, Matteo, Guido Marchetto, and Riccardo Sisto. "Push applications and dynamic content generation over content-centric networking." *International Journal of Communication Systems* 30, no. 3 (2017): e2964.

[16] Xu, Changqiao, Ping Zhang, Shijie Jia, Mu Wang, and Gabriel-Miro Muntean. "Video streaming in content-centric mobile networks: Challenges and solutions." *IEEE Wireless Communications* 24, no. 5 (2017): 157-165.

[17] Tang, Jianhua, and Tony QS Quek. "The role of cloud computing in content-centric mobile networking." *IEEE Communications Magazine* 54, no. 8 (2016): 52-59.

[18] Imbrenda, Claudio, Luca Muscariello, and Dario Rossi. "Analyzing cacheable traffic in isp access networks for micro cdn applications via content-centric networking." In *Proceedings of the 1st ACM Conference on Information-Centric Networking*, pp. 57-66. ACM, 2014.

[19] Wang, Xiaotong, Lanlan Rui, Hui Guo, and Xuesong Qiu. "Load-aware potential-based routing for the edge communication of smart grid with content-centric network." In *NOMS 2018-2018 IEEE/IFIP Network Operations and Management Symposium*, pp. 1-5. IEEE, 2018.

[20] Cai, Yueping, Jun Liu, Xinwei Fan, Ya Qiu, and Bing Tan. "Software defined status aware routing in content-centric networking." In *Information Networking (ICOIN), 2018 International Conference on*, pp. 283-288. IEEE, 2018.

[21] Chen, Jia, Bo Tong, and Hongke Zhang. "Dynamic Interest Transmission Approach for Improving Link Failure Resiliency in Content Centric Network." *IEEE Transactions on Network and Service Management* (2018).

[22] Zhang, Jianwei, Xuhui Wang, Zengyu Cai, Wanwei Huang, and Chunfeng Du. "Potential Energy Based Caching Decision Strategy for Content-Centric Network." In *Journal of Physics: Conference Series*, vol. 1087, no. 6, p. 062049. IOP Publishing, 2018.

[23] Li, Gaolei, Jianhua Li, and Jun Wu. *Fog-enabled Edge Learning for Cognitive Content-Centric Networking in 5G.* arXiv preprint arXiv:1808.09141 (2018).

[24] Lei, Kai, Shangru Zhong, Fangxing Zhu, Kuai Xu, and Haijun Zhang. "An NDN IoT Content Distribution Model With Network Coding Enhanced Forwarding Strategy for 5G." *IEEE Transactions on Industrial Informatics* 14, no. 6 (2018): 2725-2735.

[25] Li, He, Kaoru Ota, and Mianxiong Dong. "ECCN: Orchestration of Edge-Centric Computing and Content-Centric Networking in the 5G Radio Access Network." *IEEE Wireless Communications* 25, no. 3 (2018): 88-93.

In: Advanced Security Issues …
Editor: Vijey Thayananthan
ISBN: 978-1-53615-538-9

Chapter 9

SECURITY AND PRIVACY IN LOCATION-BASED SERVICES FOR 5G MOBILE COMMUNICATION

Abdullah Albelaihy[1] and Jonathan Cazalas[2]
[1]Department of Computer Science,
King Abdulaziz University, Jeddah, KSA
[2]Department of Computer Science, Florida Southern College,
Lakeland, Florida, US

ABSTRACT

With GPS-enabled devices and data connectivity now ubiquitous, Location Based Services (LBSs) have seemingly penetrated all aspects of our lives. While profoundly valuable, these services expose the user to a litany of privacy and security issues. LBS users must reveal their location, and at times other sensitive personal information, in order to effectively use the service and receive accurate results. This chapter presents the current trends of privacy techniques employed in protecting the location privacy of users in LBSs. The chapter further highlights the LBS for 5G mobile communication briefly. The outcome of this study is a taxonomy of current privacy and security techniques, which will assist researchers and developers as they look to protect the sensitive information of their customer base further.

Keywords: security, Privacy, LBS, 5G mobile communication, bloom filter

INTRODUCTION

Services need to be delivered securely from one location to another location or one location to multiple locations. When people or users operate their devices from fixed places, handling LBS is simple with minimum requirements. If people and users are moving, place or location of services will be changing with many parameters and environmental conditions. Despite many services, LBS becomes popular because the devices based on the Internet of Things (IoT) and sensors depend on many LBS. A growing number of users and devices expect to have many services simultaneously from many directions where IoT devices located. Although the goal of IoT is to handle the network communication services efficiently, LBS not only depends on the parameters and conditions around the locations but also the communication channel between the sender and receiver locations.

According to [1], users' privacy needs strong protection against LBS with maximum efficiency. Thus, authors developed a framework for protecting the privacy using redundant point-of-interest (POI) records and increasing the efficiency. Although LBS uses the proxy as a semi-trusted third party, privacy of the proxy needs protection. Here, the blind filter protocol allows the proxy to filter out redundant POI records.

Leveraging spatial diversity for privacy-aware LBS is considered in Mobile Networks [3]. Users are still struggling to manage location privacy in their daily life when they are communicating through mobile devices. Despite many protection mechanisms of location privacy, spatial diversity allows users to enhance privacy.

In privacy protection, mobile phone users can utilize the entropy-based matrices. Despite many methods for ensuring privacy protection, Voronoi graph partition can be used as another method which provides not only privacy protection but also other facilities such as query sending and k-anonymity [7].

According to [8], a novel attack on spatial cloaking schemes in LBS is introduced to protect the privacy issues of mobile users. In this proposed scheme, the short-range communication-based spatial cloaking algorithm is modified to prevent the variance based attack.

A unified framework for authenticating privacy preserving LBS is introduced in [9]. It provides the necessary authentication to queries formed from various locations and ranges. Further, the authors claimed that this framework reduces storage overheads and provides better performance than the existing techniques.

STATE OF THE ART

Privacy with Bloom Filter

According to [11-15], Bloom filter dominates in LBS for protecting the privacy and securing the users' private information. When users are in an unidentified location, the first user should identify the query's authenticity. Thus, Bloom Filter Oblivious Transfer (BLOT) is considered a novel phase privacy preserving framework. Although LBS needs such a framework for improving the privacy issues, choosing and creating correct BLOT approach depend on the Bloom filter hash function and the oblivious transfer protocol. The authors adopted Bloom Filter for the aim of securing sensitive data compared to other existing methods; unfortunately, an observation of using the bloom filter is that there is an increase in false positives, which we improve by reducing the information leakage by making the hash functions independent of each other. Further, even if the information is leaked, it will be neglected because of each of the orthogonal characteristics of each hash function; any leaked information cannot be correlated to other hash functions and can be safely ignored. Unlike the other methods which neglect this aspect, this improvement alone would achieve an increased privacy level without having to incorporate additional encryption technologies. Figure 1 illustrates the basic idea of the LBS system for preserving privacy and securing personal location data.

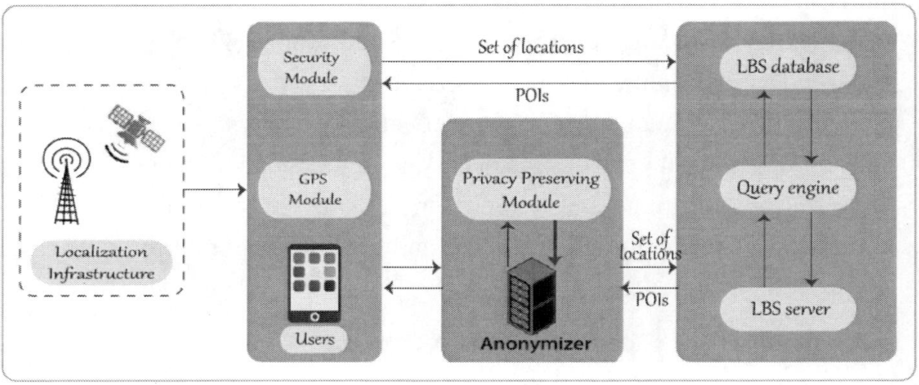

Figure 1. Structure of the LBS system [10].

Security and LBS Issues of 5G Wireless Communication

In recent years, the smartphones offer many services including LBS which is one of the features in the 5G wireless communication development. The 5G based fog computing can be used in LBS applications. For instance, fog computing is considered in airports only guide to see the direction of the specific gate and its route plan as navigation services.

According to [2], 5G-based Vehicular Social Networks (VSNs) looking for advanced LBS (ALBS). Autonomous vehicles and flying objects such as drones in the future will demand trajectory privacy-preserving scheme for maintaining ALBS in 5G or 5G+ environments. Here, authors designed a 5G-based VSN framework that incorporates Mobile Femtocell (MFemtocell) technology for providing ALBS. In addition, the authors proposed a Dynamic Group Division algorithm, which enhances the ALBS in 5G environments where the real-time demands of VSN and dynamic conditions will depend on the ALBS and intelligent ALBS.

CONCEPTS OF LOCATION PRIVACY IN LBS

Security and privacy are two different concepts, but they can interact with each other in many ways. For instance, trust can be used in both privacy and security concepts of LBS. Privacy is a concept influenced to users' personal information specially the private and confidential information should be handled privately. This prevents unauthorized accessing, tracing queries in LBS, downloading personal information without permission, etc.

The Basic Concepts of Privacy

In this chapter, we focus on introducing the notion of interactive hashing as a cryptographic primitive while differentiating it from the specifics of the implementations it may present. In this regard, this chapter shows application-independent information ideal conditions that must be ideally satisfied by the interactive hashing. Moreover, the Author provides in detail an analysis of a standard implementation of interactive hashing that meets all the conditions of our definition. From the analysis, get that it represents improvements in restricted contents than in previous attempts. In spite of its generality, the interactive hashing offers a simpler proof of security, which establishes security from a dishonest sender hence reducing his or her probability of cheating.

To prove that a sender who tries to manipulate the protocol in a way that makes it possible for output strings to have a uniquely desirable property would represent a fraction of all the strings from which the probability of both outputs will be from the set.

Also, it showed the power of interactive hashing as a cryptogrammic tool by looking into the protocols achieving oblivious transfer and which characteristically depends heavily on interactive hashing

Research Goals on Location Privacy in LBS

Following research, goals are leading us to understand the concepts of location privacy in LBS.

Privacy-Preserving: Service providers of LSB should make sure that they should not learn the queries handled between the individual user and LBS. In case, any proxy server/s is/are involved in the proposed model; the service provider should assure that proxy should not learn any queries from both LBS and users.

Communication efficiency: The algorithms and protocols used in the proposed model should be efficient in the sense that channel quality depended on the leakage of the queries should be maintained to handle the perfect communication. Further, poor channel quality increases the overhead and reduces the entropy which affects the overall communication efficiency.

Computation efficiency: Designer of the proposed model should make sure that the complexity of the protocol and algorithm should be minimum. Overall computation overhead of the proposed model depends on the complexity.

DATA ANONYMIZATION

Location cloaking provides k-anonymity in LBS and information which need to be protected. Anonymization means that identification of the data is not clear or identifiable within the set of searching area or location. In the context of 5G based LBS, location details such as address and other contact information (phone number) for preserving location privacy depend on the k-anonymity which can be referred to as k-anonymous usage of location information.

An Independent Secured Hash Function (ISHF)

Creating an appropriate hash function will not only enhance security solutions based on the Bloom filter but also protect the privacy against the

LBS. Employing ISHF allows service providers to prevent or limit information leakage. Developing and implementing Bloom filter hash function in LBS applications enable users to identify the query's authenticity. Despite many identification approaches, ISHF plays an important role in LBS applications. Thus, ISHF can be developed from Bloom filter with the following 2 considerations; they are careful to the design of hash function and minimum leak which is one-bit information per query.

Methodology and Operations

Table 1. Security issues and descriptions of some techniques used in the LBS

Techniques	Description	Security issues
BC	It allows us to secure the LBS and communications with minimum cost Also; it can merge with IoT and 5G based IoT technologies	Authentication, availability, and authorization (AAA).
OT	It creates a data transfer protocol between an LBS client and an LBS server. It depends heavily on interactive hashing	Privacy and eavesdropping
BF	It is used for hiding sensitive data, but it allows the leakage and needs a Hash function with efficient design	Privacy
BLOT	Although it does not provide the perfect protection, its privacy performance is better than the existing techniques	Eavesdropping
BC-BLOT	It enhances the not only privacy but also security issues with the 5G based encryption	AAA and malicious attack

Anonymization techniques (AT) employed as both k-anonymity and l-diversity methods to anonymize the data of the users' locations can be considered as one of the methodologies. In these methods, operations of perturbation, suppression, and generalization allow us to anonymize the data of LBS. Despite many methodologies and operations used in AT, we considered the anonymization algorithm and techniques [17-19]. In the 5G environment, blockchain (BC) concept can be used as a future methodology for improving LBS and its operation. Table 1 provides the details of Oblivious Transfer (OT), Bloom Filters (BF), BLOT and BC-based BLOT. Here, BLOT means that combination BF and OT is considered as a proposed approach. Also, BC-BLOT is one of the future frameworks combined with BC and BLOT.

Operations based on privacy breach and data utility are considered with AT for anonymizing the personal trajectory data.

Table 2. Simulation parameters

S number	Parameter	Value
1	Number of elements in bloom filter	1000
2	Bloom filter size	128 bits
3	Number of hash functions	3

HASH FUNCTIONS SIMULATION PARAMETERS

As shown in Table 2, simulation parameters allow us to conduct the simulation.

Privacy-Preserving data exchange algorithms are an important concept in the LBS analysis. As shown in figure 2, we employed the proposed data flow algorithm which allows every mobile user within the mobile user scheme of LBS region to record and exchange the user's information.

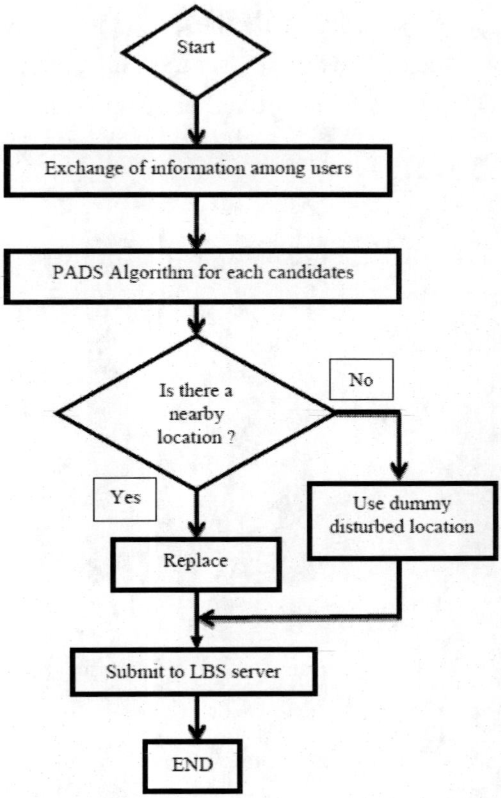

Figure 2. Proposed data flow algorithm.

Results and Implications

When users consider deploying LBS services, anonymity trajectory protection is very important to secure the LBS. If the predicted path is not interrupted by other users, the actual user's trajectory can be located. As shown in figure 3, BLOT is better than Bmobishare because they depend on the involvements of the Hash functions. When they involve more, they lose their performance.

As shown in figure 4, the entropy of BLOT (proposed approach) is better than the entropy of Bmobishare. In this comparison, 960 bits per simulation were exchanged in BLOT, which is much better than Bmobishare. Further,

BLOT provides better security with the AES256 encryption algorithm. When the entropy value of BLOT is closer to 1, information leakage will be less. Further, BLOT provides lower overhead and higher performance than the Bmobishare.

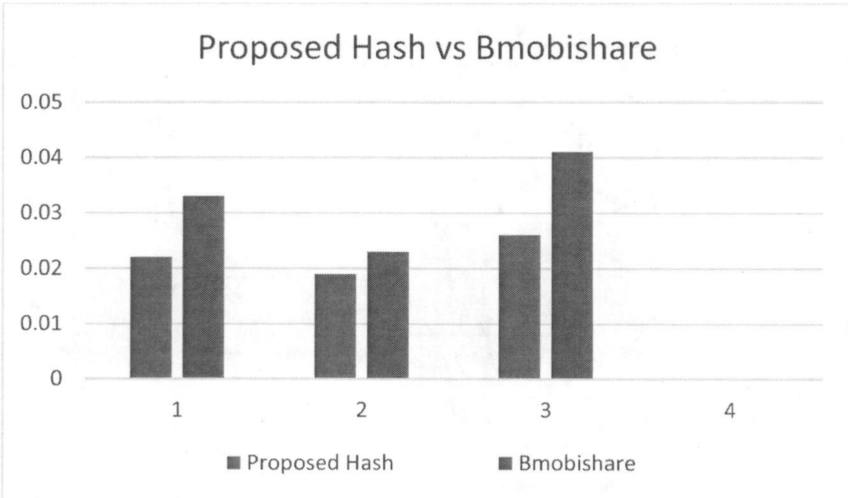

Figure 3. Hash Functions Comparison results.

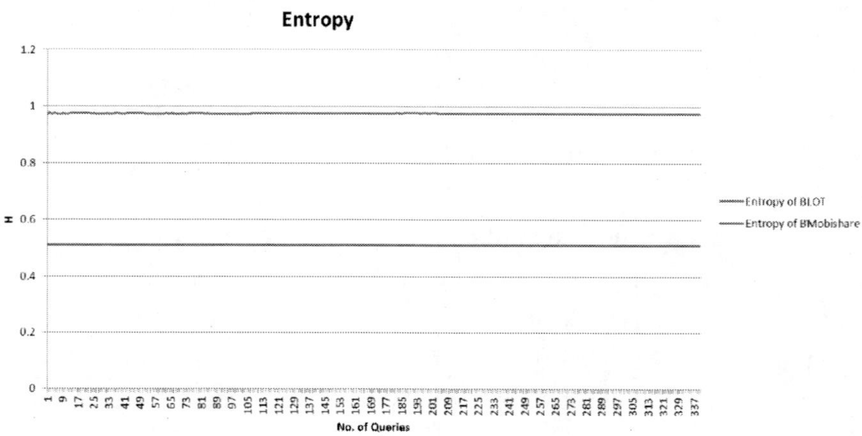

Figure 4. Performance entropy comparison.

Location Trajectory Anonymization

Although many trusted entities, anonymization in LBS must be handled very carefully according to the level of the trust used in the network. In the privacy metric, the anonymization success rate can be considered.

Although LBS faces many attacks, we have illustrated a few attacks they are trajectory and reconstruction attacks. When an attacker gets chanced to see the LBS information, the trajectory and users' location privacy will be leaked.

Trajectory attack: When LBS is unable to perform between the user and server of the LBS, the user should assume that published trajectory is being attacked. This type of attack is called a trajectory attack. It may still happen even the user is identified and removed from the trajectory.

Reconstruction attack: Although the location privacy scheme is used to hide the identity of the user, reconstruction is possible when attackers know the trajectory. The attacker tries to reconstruct the actual trajectory of the user in many ways such as assigning probabilities of possible events related to that trajectories. When the trajectory is distorted or blurred, finding the user's actual trajectory will be difficult, and the user will have location privacy.

Open Research Problems

One of the major challenges in LBS is the user's location when the user is crossing from licensed band to unlicensed band or vice versa. Finding the effectiveness of location privacy in these bands might be one of the open problems in 5G environments. The trade-off between security and privacy depends on customers' or clients' decisions which may change according to time and situations. So, decision making of security and privacy should be flexible with future LBS design and technology.

Regarding the future driverless vehicles, LBS should be dynamic and flexible with decision making and intelligent features. In this application, the collection of location and navigation data will be varying with smart facilities of smart transportation which depend on the LBS.

Sensitive location data need maximum protection, but the trade-off between security and privacy will be a challenging problem to enhance the protection.

International Mobile Subscriber Identity (IMSI) influences to LBS and provides the detection facilities when users employ the IMSI method within 5G or other existing cellular networks. To identify the specific user, IMSI detection approaches provides necessary facilities through the user's privacy [16]. Although IMSI is used for the detecting identification of the user, attackers are able to track the physical location of the users. Providing maximum privacy will be one of the challenges when IMSI is employed in the detection.

According to [4], a proposed solution based on the cloaked region construction algorithm (CRCA) provides efficient privacy-preserving LBS. Query similarity attack is one of the problems in LBS, but Dice's coefficient approach can be used to avoid this attack. Further, CRCA improves the quality of the query services which not only protect the users' LBS but also it provides a high degree of privacy.

Applications of LBS [5] are increasing from social networking to scientific and evolving communication such as IoT, 5G, 5G+, etc. Although many of these applications depend on the distances between the users/devices, security issues based on the distance briefly mentioned using localization and proximity (LP).

Achieving load balancing between privacy protection level and power consumption in LBS is possible through the model given in [6]. In this model, authors have introduced 6 factors they are backlight, CPU, WiFi, memory, bandwidth, and GPS which allow us to analyze the behavior of LBS applications.

CONCLUSION

We investigated the ways of applying hash functions for increasing entropy which means hash functions reduces the information leakage. In this research, BLOT mechanism allows addressing the security deficiencies in the LBS applications through the 2 combined protocols which are oblivious transfer and with Bloom filters. Also, BLOT enhances the security of LBS and its communication when attacks happen between the client and the server locations. Performance entropy based on BLOT results better than the existing approaches such as BMobishare.

We studied security and privacy in LBS for vehicular and 5G mobile communications. Based on the analysis of BLOT, some future challenges of LBS are considered as open research problems.

REFERENCES

[1] Chen, Jing, Kun He, Quan Yuan, Min Chen, Ruiying Du, and Yang Xiang. "Blind Filtering at Third Parties: An Efficient Privacy-Preserving Framework for Location-Based Services." *IEEE Transactions on Mobile Computing* (2018).

[2] Liao, Dan, Hui Li, Gang Sun, Ming Zhang, and Victor Chang. "Location and trajectory privacy preservation in 5G-Enabled vehicle social network services." *Journal of Network and Computer Applications* 110 (2018): 108-118.

[3] He, Xiaofan, Richeng Jin, and Huaiyu Dai. "Leveraging Spatial Diversity for Privacy-Aware Location-Based Services in Mobile Networks." *IEEE Transactions on Information Forensics and Security* 13, no. 6 (2018): 1524-1534.

[4] Saravanan, P. Shanthi, and S. R. Balasundaram. "Protecting Privacy in Location-Based Services Through Location Anonymization Using Cloaking Algorithms Based on Connected Components." *Wireless Personal Communications* 102, no. 1 (2018): 449-471.

[5] Usman, Muhammad, Muhammad Rizwan Asghar, Imran Shafique Ansari, Fabrizio Granelli, and Khalid A. Qaraqe. "Technologies and Solutions for Location-Based Services in Smart Cities: Past, Present, and Future." *IEEE Access* 6 (2018): 22240-22248.

[6] Alrahhal, Mohamad Shady, Maher Khemekhem, and Kamal Jambi. "Achieving load balancing between privacy protection level and power consumption in location-based services." *International Research Journal of Engineering and Technology (IRJET)* 11, no. 3 (2018).

[7] Ma, Chunguang, Changli Zhou, and Songtao Yang. "A voronoi-based location privacy-preserving method for the continuous query in LBS." *International Journal of Distributed Sensor Networks* 11, no. 3 (2015): 326953.

[8] Niu, Ben, Xiaoyan Zhu, Qinghua Li, Jie Chen, and Hui Li. "A novel attack to spatial cloaking schemes in location-based services." *Future Generation Computer Systems* 49 (2015): 125-132.

[9] Hashem, Tanzima, Shudip Datta, Tanzir Ul Islam, Mohammed Eunus Ali, Lars Kulik, and Egemen Tanin. "A unified framework for authenticating privacy-preserving location-based services." In *Second International ACM Workshop on Managing and Mining Enriched Geo-Spatial Data*, pp. 13-18. ACM, 2015.

[10] Sun, Yanming, Min Chen, Long Hu, Yongfeng Qian, and Mohammad Mehedi Hassan. "ASA: Against statistical attacks for privacy-aware users in Location Based Service." *Future Generation Computer Systems* 70 (2017): 48-58.

[11] Albelaihy, Abdullah, Jonathan Cazalas, and Vijey Thayananthan. "A Survey: The Current Trends of Privacy Techniques for Protecting the Location Privacy of Users in LBS." *Journal of Theoretical & Applied Information Technology* 96, no. 15 (2018).

[12] Albelaihy, Abdullah, and Jonathan Cazalas. "A survey of the current trends of privacy techniques employed in protecting the Location privacy of users in LBSs." In *Anti-Cyber Crimes (ICACC), 2017 2nd International Conference on*, pp. 19-24. IEEE, 2017.

[13] Albelaihy, Abdullah, Jonathan Cazalas, and Vijey Thayananthan. "Privacy-Preserving Queries for LBS: Independent Secured Hash Function." *Journal of Theoretical & Applied Information Technology* 96, no. 11 (2018).

[14] Albelaihy, Abdullah, and Jonathan Cazalas. "Privacy-Preserving Queries for LBS: Hash function secured (HFS)." In *Anti-Cyber Crimes (ICACC), 2017 2nd International Conference on*, pp. 7-12. IEEE, 2017.

[15] Albelaihy, Abdullah, Jonathan Cazalas, and Vijey Thayananthan. "BLOT: A Novel Phase Privacy Preserving Framework for Location-Based Services." *International Journal of Advanced Computer Science and Applications* 9, no. 5 (2018): 95-104.

[16] Baek, Jaejong, Sukwha Kyung, Haehyun Cho, Ziming Zhao, Yan Shoshitaishvili, Adam Doupé, and Gail-Joon Ahn. "Wi Not Calling: Practical Privacy and Availability Attacks in Wi-Fi Calling." In *Proceedings of the 34th Annual Computer Security Applications Conference*, pp. 278-288. ACM, 2018.

[17] Pingley, Aniket, Wei Yu, Nan Zhang, Xinwen Fu, and Wei Zhao. "A context-aware scheme for privacy-preserving location-based services." *Computer Networks* 56, no. 11 (2012): 2551-2568.

[18] Hasan, A. S. M., Qiang Qu, Chengming Li, Lifei Chen, and Qingshan Jiang. "An Effective Privacy Architecture to Preserve User Trajectories in Reward-Based LBS Applications." *ISPRS International Journal of Geo-Information* 7, no. 2 (2018): 53.

[19] Saravanan, P. Shanthi, and S. R. Balasundaram. "Protecting Privacy in Location-Based Services Through Location Anonymization Using Cloaking Algorithms Based on Connected Components." *Wireless Personal Communications* 102, no. 1 (2018): 449-471.

In: Advanced Security Issues …
Editor: Vijey Thayananthan
ISBN: 978-1-53615-538-9

Chapter 10

LEARNING STYLE TRANSFORMATION USING INTELLIGENT IOT, 5G AND BEYOND

***Vijey Thayananthan*[1] *and Benachour Phillip*[2]**
[1]Department of Computer Science,
King Abdulaziz University, Jeddah, KSA
[2]Department of Computer and Communication,
Lancaster University, UK

ABSTRACT

All learning styles depend on evolving and merging technology which transforms the users' most of the current and existing services into IoT based 5G scheme. Learning style transformations for managing high-tech strategies such as intelligence, automation, etc. depend on the cyber-physical system (CPS). Although current technologies provide many facilities, learning style transformation increases the costs of public services with emerging technologies such as IoT based 5G. Reducing the costs of educational services is one of the challenging problems in industry 4.0. To improve the learning style transformation based on the emerging technologies, the method we introduced in this chapter will minimize the cost which depends on the energy saving (ES) and consumptions. As a result, we provide the theoretical model which helps users to indicate the

ES when we employ the intelligent CPS and IoT based 5G technology. Finding solutions for improving learning style transformation with minimum cost will be the conclusion of future educational services.

Keywords: intelligent, CPS, Industry 4.0, learning style transformation

INTRODUCTION

Improving learning style transformation should help to increase the facilities with the available technology. A scientist invented many innovations for improving virtual learning and medical facilities with modern electronic technologies, but still, medical cost is very high. Industry 4.0 will find the compromised solutions which not only reduce the costs of all compulsory public services but also increase the quality of the services with the evolving and merging technologies. Despite many intelligent systems, this chapter deploys the CPS with the ES approaches which allow users to increase the lifetime of the devices employed for learning activities. Regarding the automation services in learning environments, the error rate may be slightly lower than the existing online educational systems. Here, the research question is about ES and intelligent CPS which provides the comporimization in public services. Can we reduce the overall costs of public services using minimum energy and maximum protection?

The main purpose of this chapter is to develop a theoretical model of ES system using intelligent CPS which not only reduce the unnecessary delays during the online academic services but also increase the quality of the online services that enhance the livability of industry 4.0. Although the main focus considers the learning style transformation with efficient solutions, this chapter introduces the dynamic and intelligent approaches of the future solutions to detect the evolving technical problems related to the emerging technologies.

As a problem statement, this chapter considers the minimization of public costs related to public service problems depended on many factors including intelligent issues and ES. To improve the learning style transformation problems, all public services used within the IoT industry 4.0 should be monitored intelligently. Despite many monitoring facilities and intelligent activities, optimization of accurate network management during the online services and automation facilities of the learning style transformation can be considered with the IoT based 5G infrastructure.

Although the application layer is responsible for setting many different applications of the public services, it provides efficient management through the existing protocols. Thus, learning style transformation using intelligent IoT and 5G+ technologies will encourage service providers to manage the access control of online learning facilities. Therefore, introducing intelligent protocols will enhance the features of access control which protects the application layer. Despite many intelligent protocols and algorithms, intelligent activities of access control should be considered to improve the ES management in public services.

LITERATURE REVIEW AND RELATED WORK

According to [1], industry 4.0 somehow influences to CPS for improving daily learning activities which people face in the public and private sectors. Although CPS is dominating in industry 4.0, big data plays an important role to change the learning style in many different ways. For instance industry 4.0 based automotive system is the revolution of the existing automobile system using big data for analyzing daily activities of transportation services. Transformation of manufacturing [2] is evolving with the technology which simplifies the daily working patterns and process. As far as [3, 20] is concerned, smart manufacturing not only increases the profits but also improves the conditions of working style which transforms the learning style in industry 4.0. The 5G is a key enabling technology to revolutionize the future of industrial Cyber-physical IoT system (CPIoTS) [4]. Despite the daily services involved with IoT, CPIoTS provides support to enhance the interoperability and interaction of various machines used in the learning style transformation. According to [5], industrial cyber-physical systems (ICPSs) is emerging with the legacy production systems. Although existing solutions of ICPS supports the remote configurations and operations of productions, malicious attacks not only damage the resources used in the ICPS but also threaten the human lives. In industry 4.0, learning style transformation depends on the automation, intelligent, etc. which provide systematic learning with intelligent ways. Learning styles are not only changing the personal appearance but also improving the quality of learning with evolving and merging technologies. Despite many energy-efficient technologies, the automation system needs maximum protection with a minimum cost which depends on energy management. The future automation system expects to have dynamic power generated through energy harvesting (EH). In this state of the art, CPS plays an important role in connecting physical nodes known as mobile devices and energy source which may be in random and remote locations. Here, wireless charging units considered as virtual energy sources may be employed for harvesting energy. Business processing [6] moves with people's lives and their demands according to the industry 4.0 environments. A novel IoT architecture based on 5G-IoT [7] will be the next generation technology for improving IoT industry 4.0 and provide merging facilities with future applications. Implying systemic transformations at micro, meso and macro levels of analysis [8] changes the learning style of industry 4.0 where products, production processes, and business models are upgrading into the new technological paradigm. Good jobs in industrial upgrading and transformation [9] is another improvement in industry 4.0 because evolving and merging new technology has changed the legacy of the existing job functions. Although industry 4.0 revolution in Russia [10] fundamentally transforms modern production, new technological achievements, including digitalization and robotization, artificial intelligence and the Internet of things (IoT), new materials and biotechnology create new opportunities to improve the economy of the industry 4.0. Smart factory depends on the requirements which are varying according to the particular country's environmental conditions [11] and their industrial solutions. Edge computing gateway of the industrial IoT using multiple collaborative microcontrollers improves industry 4.0 where minimizing response time is one of the key elements for managing network devices [12]. Based on the

[13], digital manufacturing, digital-networked manufacturing, and new-generation intelligent manufacturing allow the manufactures to follow their daily tasks systematically. According to [14], the authors proposed the model "reliability modeling and evaluation of CPS considering communication failures" using Markov process. Despite the reliability in CPS, working state, repairing state and delay-repairing state are considered for improving the protection in the industry 4.0. Regarding learning style transformation, a framework for operative and social sustainability functionalities in human-centric CPPS (Cyber-Physical Production Systems) [15] provides many advantages to human operators.

INTELLIGENT ISSUES OF CPS IN AN AUTOMATION SYSTEM

The IoT based 5G schemes are growing among the services used within the public and private organizations of industry 4.0. This scheme needs intelligent infrastructure which allows the service provider to monitor and solve the automation facilities quickly and dynamically. The CPS technology innovation Centre within the industry 4.0 will offer each program the possibility to develop applications related to smart systems [22], as follows:

- Automation: IoT, smart factory, intelligent systems, electronics for the wireless sensors system
- Manufacturing: additive manufacturing for smart industry 4.0, smart robots for implementing and producing actual products designed by the factory
- Infrastructure: smart 5G based systems, smart devices, and smart communication structures
- Software: networking and infrastructure, Big Data of intelligent schemes, Data analytics

- Energy: alternative energy, smart grid, charging facilities wirelessly and dynamically

The automation system considered has been employed to future IoT based 5G infrastructure which allows improving the industrial communications.

Figure 1 shows the SDIAN (Software Defined Industrial Automation Network) which helps future users to improve and transform the learning style in industry 4.0. This approach not only minimizes the industrial cost but also provide a systematic and manageable learning style [24].

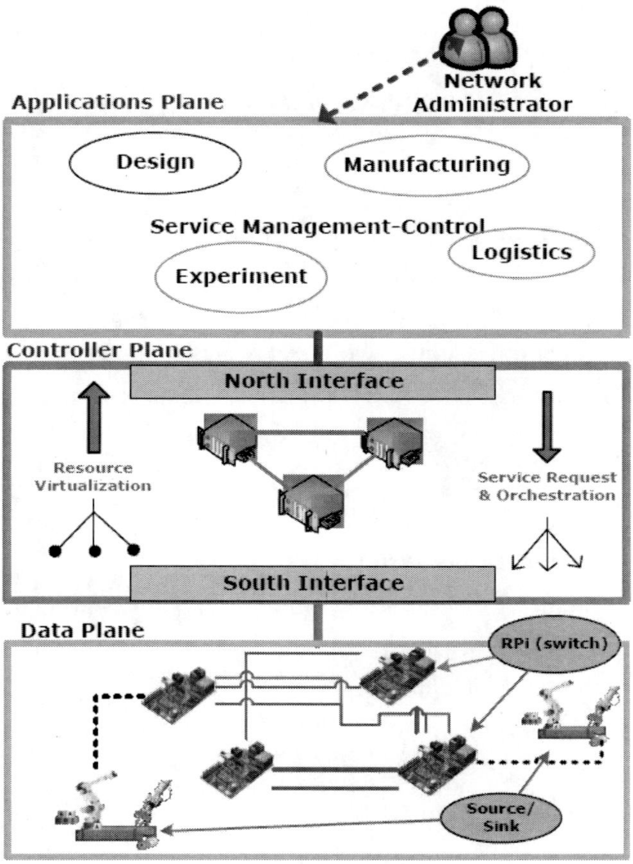

Figure 1. SDIAN communication framework [24].

System Models for Industrial CPSs

Intelligence enabling the communication and cooperation of people and machines with the systems of information and communication technology in real time.

A Theoretical Model of the Automation System

As shown in Figure 2, this chapter proposes the theoretical model which represents the simple illustration of the automation system. In this model, SDIAN provides necessary facilities to improve industrial learning in which industries transform from 4G to 5G+. Here, the legacy of the business systems may be updated without changing the major hardware.

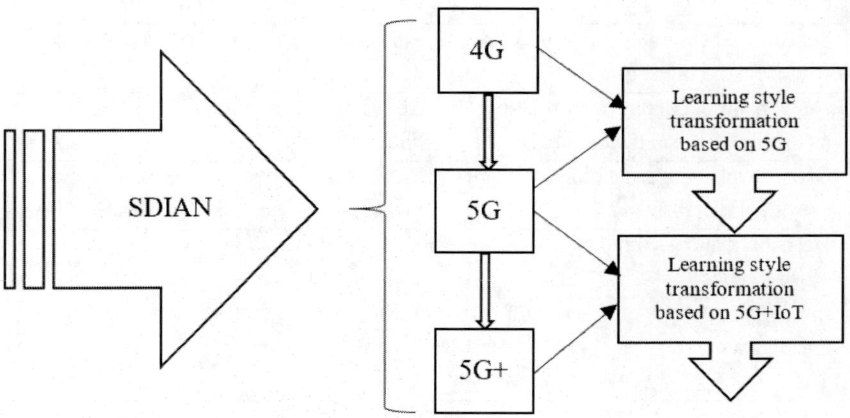

Figure 2. A theoretical model of learning style transformation in industry 4.0.

Further, this theoretical model learns the existing industrial learning style and transform the industrial programming approaches into the potential system according to the evolving technology.

The CPS and Learning Style Transformation

The idea behind a CPS is to connect the virtual and physical industry 4.0 environments where sensors used in the automation and intelligent system are able to gather all monitored data related to public services. Further, they provide useful feedback to users or all physical entities connected to IoT based 5G systems.

The sensor data gathered from the physical machine is transmitted through IoT and allows a remote operator to monitor and control the equipment based on the data received. Improvement of the learning style depends on CPS which allows the communication between machines and humans.

Table 1. Examples of a learning style transformation

CPS	Learning style transformation
Sensors and communication network with a smart connection	Plug and play with all available mobile devices
Conversion of data into information using analytic data techniques	Converting data to image or audio or video
Cyber levels with technologies which provide the improvements	Working human learning pattern, time management, etc.
Decision making with diagnostic and sensing devices	Visualization and understanding with real-time activities
Intelligent configuration with self-management	Self-configuration for varying and dynamic situations

These systems can gather and process data for transforming the output into a new format according to the evolving and merging technology.

Infrastructure and Future Automation System

Automation approaches in modern industries provide a number of benefits with minimum cost and manpower which changes the learning style in the industry 4.0. According to [16, 19], self-aware and self-configuring factories depend on automation system where industrial analytics methodology and fog computing CPS are the new method and technology respectively. Here, the legacy of the industry evolve with industry 4.0 embedded system applications and emerge with big data, artificial intelligence, and machine learning.

The CPPS Functionalities

The idea behind a CPPS is to improve the industrial control functionalities which are being evolved with new technologies and techniques. The intelligence and decision-making approaches depend on the functionalities we employ in the industry 4.0. Following CPPS functionalities defined in [15] encourage the users to improve the use of technology according to the learning style transformation. They are Silent Teacher; Knowledge Manager; Resource Integrator; Care Giver; Risk Manager; Flow Master.

ANALYSIS OF INDUSTRY 4.0 SOLUTIONS

In the learning style transformation, a theoretical model will allow us to analyze the solution with evolving technologies such as IoT based 5G and 5G+. Thus, we can enhance the industry 4.0 solutions using a theoretical model which incorporate with the automation and intelligent systems.

The Automation System in

Although many benefits discussed in the literature reviews, the use of a theoretical model and its solutions in the industry 4.0 will encourage us to minimize the cost and increases the facilities of learning style. Modern academic institutions transform the learning style into the potential system such as new automation according to current demand and requirements. In this analysis, challenges of learning style in the professional industries may be considered in the specific application of the industry 4.0. For instance, the current automotive services and future driverless vehicles are being evolved with new automation technology.

State of the art in industry 4.0 based automation system is the revolution of the existing manufacturing style system which is a legacy of the industrial framework.

Despite many industrial control systems, evolving technology, global warming and other factors such as energy harvesting, intelligent issues of minimizing energy cost, etc. will transform the learning style in industrial 4.0.

The Intelligent System in Industrial CPSs

According to the survey on intelligent control systems, learning style for managing industrial CPS transform the daily life into a systematic way with good intelligent solutions and decision making. The ICPS involved with the social activities, learning style transformations, etc. needs some additional solutions which include security [17]. The intelligent system will play an important role in the industry 5.0 which utilizes IoT but differs from predecessor automation systems by having three-dimensional (3D) symmetry in innovation ecosystem design [23].

Learning Style Transformation in Education Environments

Education environments were encouraging to learn new programs which fit to manage our daily learning. In industry 4.0, education environments are not only delivering courses but also transferring industrial programs such as the experience of real-time engineering. Industry 4.0 manufacturing and engineering education [18, 21] provide nine technological pillars in the higher education curriculum which improves the learning style in the industry 4.0.

Conclusion

The IoT based 5G concept in the automation system will enhance intelligent procedures when CPS influences in future online academic services. This chapter introduces the theoretical model that addresses the intelligent issues of CPS using IoT based 5G architecture. This model will not only the increases in the intelligent atmosphere of the public services, but also it provides the ES which reduces the overall costs of the industry 4.0. In this research, we have studied the current and future academic services. Further, the theoretical model employs the SDN to improve intelligent issues with minimum energy consumptions. The dynamic policies which we can develop for our theoretical model reduce the expected waiting time to improve the learning style transformation. In future works, we can simulate the proposed theoretical model and analyze the results with different public services.

References

[1] Xu, Li Da, and Lian Duan. "Big data for cyber-physical systems in industry 4.0: a survey." *Enterprise Information Systems,* (2018): 1-22.

[2] Al-Ali, A. R., Ragini Gupta, and Ahmad Al Nabulsi. "Cyber-physical systems role in manufacturing technologies." In *AIP Conference Proceedings*, vol. 1957, no. 1, p. 050007. AIP Publishing, 2018.

[3] Jbair, Mohammad, Bilal Ahmad, Ahmad Mus' ab H. and Robert Harrison. "Industrial cyber-physical systems: A survey for control-engineering tools." In *2018 IEEE Industrial Cyber-Physical Systems (ICPS)*, pp. 270 - 276. IEEE, 2018.

[4] Li, Song, Qiang Ni, Yanjing Sun, Geyong Min and Saba Al-Rubaye. “Energy-Efficient Resource Allocation for Industrial Cyber-Physical IoT Systems in 5G Era”. *IEEE Transactions on Industrial Informatics,* 14, no. 6 (2018): 2618 - 2628.

[5] Ulz, Thomas, Thomas Pieber, Christian Steger, Sarah Haas, and Rainer Matischek. “secured remote configuration approach for industrial cyber-physical systems.” In *2018 IEEE Industrial Cyber-Physical Systems (ICPS)*, pp. 812 - 817. IEEE, 2018.

[6] Özüdoğru, Ayşe Göksu, Esra Ergün, Djihane Ammari, and Ali Görener. “How Industry 4.0 Changes Business: A Commercial Perspective”. *International Journal of Commerce and Finance,* 4, no. 1 (2018): 84.

[7] Rahimi, Hamed, Ali Zibaeenejad and Ali Akbar Safavi. *“A Novel IoT Architecture based on 5G-IoT and Next Generation Technologies.”* arXiv preprint arXiv:1807.03065 (2018).

[8] Galli, Stefania. *Towards a New Technological Paradigm Based on Industry 4.0: Opportunities and Challenges for Innovation Policies.* PhD diss., University of Trento, 2018.

[9] Wei, Nai-Chieh, Hsiang-Chen Hsu and Ai-Ling Yang. “Good Jobs in Industrial Upgrading and Transformation.” *International Journal of Organizational Innovation (Online),* 10, no. 3 (2018): 167 - 188.

[10] Vasin, S., L. Gamidullaeva, E. Shkarupeta, A. Finogeev, and I. Palatkin. “Emerging Trends and Opportunities for Industry 4.0 Development in Russia.” *European Research Studies Journal,* 21, no. 3 (2018): 63 - 76.

[11] Odważny, Filip, Olga Szymańska and Piotr Cyplik. “Smart Factory: The Requirements for Implementation of The Industry 4.0 Solutions in Fmcg Environment-Case Study.” *LogForum,* 14, no. 2 (2018).

[12] Chen, Ching-Han, Ming-Yi Lin and Chung-Chi Liu. “Edge Computing Gateway of the Industrial Internet of Things Using Multiple Collaborative Microcontrollers.” *IEEE Network,* 32, no. 1 (2018): 24 - 32.

[13] Zhou, Ji, Peigen Li, Yanhong Zhou, Baicun Wang, Jiyuan Zang and Liu Meng. "Toward New-Generation Intelligent Manufacturing". *Engineering,* 4, no. 1 (2018): 11 - 20.

[14] Yang, Yi, Sixin Wang, Meilin Wen and Wei Xu. "Reliability Modeling and Evaluation of Cyber-Physical System (CPS) Considering Communication Failures," *Journal of the Franklin Institute,* (2018).

[15] Pinzone, Marta, Federico Albè, Davide Orlandelli, Ilaria Barletta, Cecilia Berlin, Björn Johansson and Marco Taisch. "A framework for operative and social sustainability functionalities in Human-Centric Cyber-Physical Production Systems". *Computers and Industrial Engineering,* (2018).

[16] O'Donovan, Peter. "An industrial analytics methodology and fog computing cyber-physical system for Industry 4.0 embedded machine learning applications" (2018).

[17] Ding, Derui, Qing-Long Han, Yang Xiang, Xiaohua Ge and Xian-Ming Zhang. "A survey on security control and attack detection for industrial cyber-physical systems." *Neurocomputing,* 275 (2018): 1674-1683.

[18] Hernández-Muñoz, Guadalupe Maribel, Lizbeth Habib-Mireles, Flor Araceli García-Castillo and Fernando Montemayor-Ibarra. "Industry 4.0 and Engineering Education: An Analysis of Nine Technological Pillars Inclusion in Higher Educational Curriculum". In *Best Practices in Manufacturing Processes*, pp. 525 - 543. Springer, Cham, 2019.

[19] Verba, Nandor, Kuo-Ming Chao, Jacek Lewandowski, Nazaraf Shah, Anne James and Feng Tian. "Modeling industry 4.0 based fog computing environments for application analysis and deployment". *Future Generation Computer Systems,* 91 (2019): 48 - 60.

[20] Zheng, Pai, Zhiqian Sang, Ray Y. Zhong, Yongkui Liu, Chao Liu, Khamdi Mubarok, Shiqiang Yu and Xun Xu. "Smart manufacturing systems for Industry 4.0: Conceptual framework, scenarios, and future perspectives". *Frontiers of Mechanical Engineering,* (2018): 1-14.

[21] Paravizo, Esdras, Omar Cheidde Chaim, Daniel Braatz, Bernd Muschard and Henrique Rozenfeld. "Exploring gamification to

support manufacturing education on industry 4.0 as an enabler for innovation and sustainability". *Procedia Manufacturing,* 21 (2018): 438 - 445.

[22] Centea, D., Singh, I., and Elbestawi, M. (2018). Framework for the Development of a Cyber-Physical Systems Learning Centre. In *Online Engineering and Internet of Things*, (pp. 919 - 930). Springer, Cham.

[23] Özdemir, Vural and Nezih Hekim. "Birth of industry 5.0: Making sense of big data with artificial intelligence,"the internet of things" and next-generation technology policy". *Omics: A Journal of Integrative Biology,* 22, no. 1 (2018): 65 - 76.

[24] Ahmed, Khandakar, Jan Blech, Mark Gregory and Heinz Schmidt. "Software-Defined Networks in Industrial Automation." *Journal of Sensor and Actuator Networks,* 7, no. 3 (2018): 33.

ABOUT THE EDITOR

Vijey Thayananthan, PhD is an Associate Professor at Computer Science Department in the King Abdulaziz University, Jeddah, KSA. He obtained his PhD in Engineering (Digital communication engineering) from Department of Communication Systems, University of Lancaster, UK, in 1998. Further, he had worked as Postdoctoral research Fellow in Department of Computer Science, Glasgow/Sterling University and Communication division, Department of Electrical and Electronic Engineering, Strathclyde university, UK for 2 years. Since 2000, he had been working as a Research engineer and senior algorithm development engineer in Advantech Ltd, Southampton University Science Park, UK and Amfax Ltd, UK respectively. His research interests include wireless communication algorithm design and mobile communication analysis, security management of communication network and big data, computer security and wireless sensor network. He has been a full-time member of the Institution of Engineering Technology (IET), UK since 2005.

INDEX

B

C

D

E

F

G

H

I

K

L

M

Q

R

S

T

U

V

W

Related Nova Publications

IoT: Platforms, Connectivity, Applications and Services

Author: Abdulrahman Yarali

Series: Computer Science, Technology and Applications

Book Description: In this book, there are 16 chapters which cover a broad range of topics such as platforms, technologies, generating business value, delivering smart, sustainable energy solutions, smart communities and citizens, manufacturing, healthcare, security and privacy, commercial drones and many other related IoT topics.

Hardcover ISBN: 978-1-53613-400-1
Retail Price: $230

The Internet of Things (IoT): Applications, Technology, and Privacy Issues

Author: Silvia Watts

Series: Internet Theory, Technology and Applications

Book Description: This book examines architectures, technologies and applications in the Internet of Things. It presents the development and implementation of an industrial communication system for remote monitoring and management of industrial processes as well as the current IoT governance scenario, its main challenges, the role of the governments and the need to develop public-private partnerships for technical cooperation in the governance of IoT resources.

Hardcover ISBN: 978-1-63484-626-4
Retail Price: $110

To see complete list of Nova publications, please visit our website at www.novapublishers.com